DISCOVERING THE UNIVERSE

DISCOVERING
the
UNIVERSE

A History of
ASTRONOMY

COLIN A. RONAN

Basic Books, Inc., Publishers

NEW YORK

Library of Congress Catalog Card Number: 72–135556
SBN 465–01670–7
Manufactured in the United States of America

Contents

List of Illustrations

DISCOVERING THE UNIVERSE

Introduction

Man's struggle to understand the world around him stretches back into the uncharted regions of human history; it is as old as man himself. Man has always been curious about how things happen and why things are there. The growing body of knowledge that he has slowly and painstakingly gathered makes a fascinating story, for it is nothing less than the saga of the development of human thought. And throughout it, like an unbroken thread, is his bold attempt to inquire into the nature of the stars and the universe that contains them. This inquiry has been continuous, partly because the subject itself has so intrigued man and partly because its practical uses have held his attention.

At a first glance it may not seem that a study of the stars can have any practical effect, certainly on primitive man, but a moment's thought will show that this is far from the case. The very determination of time, of the seasons, of a calendar is dependent on the movement of the heavens—or on the motion of the earth in its annual orbit round the sun, as we would put it today. And if we want to find our position on the earth's surface, to survey boundaries between one country and another, in fact to know precisely where we are, celestial measurement is the only way of doing so. Studies of the stars also provide knowledge that is difficult, expensive, and sometimes impossible to get in a laboratory, for stars are immensely hot bodies whose conditions we cannot fully simulate without annihilating ourselves.

In past times there was another very powerful reason for examining the heavens, one that overruled almost every other consideration; this was the belief that the stars swayed the destinies of rulers and those who lived under them. Today this belief is seen to have no scientific basis, no foundation in reality, and is no more than a harmless superstition, but in earlier times when knowledge of the universe was more restricted and there was no hard and fast line between science and the supernatural world of spirits, gods, and magic influences, it seemed logical enough. The most learned and enlightened men accepted it, and the study of astrology was considered an honorable calling. Astrology entered into every walk of life. Leaders consulted the stars before they made any public decision, particularly about war; as soon as a child was born, the aspects of the stars at the time of his birth were noted and their influence calculated to provide a horoscope that would guide his future life. And when health failed, treatment had to be given at auspicious times that were chosen with reference to the heavens, and even doctors' herbal and chemical remedies were only effective if compounded at the correct astrological times.

Astrology also exerted a powerful influence on early chemistry, or alchemy. Chemical reactions were also thought to be influenced by the aspects of the stars, and all important alchemical transformations, particularly those concerned with manufacturing an elixir to bring eternal youth, or compounding the "philosopher's stone" that would turn common metals into gold, were carried out only at appropriate times determined from the appearance of the heavens. Yet though this was all superstition, there was one partly scientific aspect of astrology—natural astrology—though the reasons why this was so escaped those who practiced it. Natural astrology was concerned with the behavior of nature and covered such things as the weather, the seasons, and the growth of crops.

The weather was an obvious case for astrological influence; many things that we would call astronomical, such as meteors ("shooting stars") and comets, were then thought to be meteorological. But the element of superstition was still present; halos around the sun, rings around the moon, and such catastrophes as earthquakes, floods, and drought were all mistakenly attributed to celestial influences. With the seasons, natural astrology was on much firmer ground, for these depend on the position of the earth in its orbit around the sun, which causes different stars to be visible at night. As far as the question of crop growth goes, this is partly seasonal but not completely so; there is a link, for instance, between tree growth and radiation from the gases that compose the sun. There is even a possible connection between the time crops are sown and the phases of the moon; the relationship is uncertain and even though some gardeners are sure of it, the scientist is not yet convinced, in spite of his knowledge that the moon causes small "tides" in the earth's crust as well as tides in the sea. At all events, this natural astrological influence was not so far-fetched, even by modern standards. But in this book we shall not primarily concern ourselves with astrology, but with the history of man's discovery of the nature of the universe.

This is a large subject, and there is more than one way in which it can be tackled. We can begin at the beginning and continue to today, mentioning each discovery and theory as it occurred. But the story is complex and at any time there were a number of different things happening at once; some would study the stars, others would be concerned with navigation or calendar-making, a few might be designing and building instruments. If we try to go through everything as it happened and make no selection, we shall end up with a catalog, or a muddle. The only satisfactory way—at least after concerning ourselves with very primitive times—is to deal with it subject by subject, and this is what has been done in the pages that follow. At the end of the book we shall attempt the exciting task of using this knowledge of history to see if we can glimpse where the astronomer is going and assess how good—or bad—some modern theories are.

But first, we must delve into the dim shades of the past and put ourselves in the place of those who began astronomical science.

1 ✳ Early Beliefs

Astronomy, the study of the heavens, is the oldest branch of science, and its beginnings go back long before recorded history. To learn of the earliest beliefs we must think about ancient traditions and examine the legends that were passed by word of mouth from father to son. We must also use our imagination, trying to forget our present knowledge, and put ourselves in the place of primitive man.

To begin with, we must consider what would have impressed us about the sky if we had lived 6,000 years ago, or more. Probably the most obvious event would have been the way in which night follows day. The perpetual change of night into day and day into night regulated the lives of people living close to nature. Artificial lighting, when at last invented, was inconvenient and uncommon, so it was generally a case of starting the day at sunrise and going to bed at sunset. The day, the time of light, was obviously governed by the sun, for when it was not present in the sky, there was darkness. The sun, with its powers of light and heat, was clearly of the greatest importance, and we should not be surprised to realize that, at some very early time, it began to be worshipped as a god—a practice that continues in some primitive societies today. But the sun did not stand still. At sunrise it began to peep over the horizon, and then, as the day wore on, it rose high in the sky. Later it would sink down toward the ground and disappear again. And the horizon it rose from was opposite the horizon in which it sank; in its rising and falling movement the sun moved across the sky from one side of the world to the other—from what we call the east to the west.

The sun's movement was regular, happening every day and never seeming to alter from one generation to another; it was the same as far back as tribal memory could go. So one conclusion we could come to was that the sun was eternal. We could also see that its regular motion might be used for determining time, because time depended on its position in the sky. When the sun had reached its greatest height, it was halfway between its moment of rising and moment of setting, so this was midday, or noon. The sun cast shadows, which changed during its daily movement. These shadows diminished as the sun rose in the sky and lengthened as it set, changing their direction as the sun moved from the east to the west. A shadow time indicator, such as a tree or a stick driven into the ground, would show this, and it must have been devised very early.

When the sun had set in the west, darkness fell, and continued until the sun appeared on the eastern horizon at dawn. Where had it gone during the night? This is an obvious question and must have exercised the minds of the earliest thinkers. Today we can provide a very sophisticated answer, talking about a spinning earth shaped like a globe, but in primitive times there was no reason to consider the earth as round—indeed every shred of evidence seemed to point to its being flat—and any idea of its movement would appear not only wrong but stupid. The earth is huge and still; it does not tremble or shudder all the time as one would expect if it moved, so it is clear that it is fixed quite firmly—literally, as still as a rock. In fact, it is probable that 6,000 or more years ago no one had even thought for a moment of the idea of a moving earth.

So let us try to imagine what conclusions we could make about the movement of the sun during the night with the facts at our disposal. We have a solid earth, fixed and unmovable, and we have concluded that it is flat. Of course, we know from our everyday experience that it is not completely flat—there are hills, valleys, and mountains—but if you look from a great distance, even mountainous country seems to stretch in a level sweep into the distance and, of course, however high the mountains seem to be, they rest on flat ground. What of the sun? It rises above the flat earth at one side, goes up into the sky, and comes down again on the other side; then it goes back again so that it may repeat the same cycle of changes again and again. At night it does not come above the horizon; otherwise it would be bright instead of dark. Clearly, the sun must go down below the earth at night and move from west back to east. Precisely how it moves underneath the earth we have no way of finding out. Perhaps it travels under the edge of the flat earth or, possibly, it goes in a direct line from west to east; certainly it must do one or the other of these things. We have, then, a simple theory of the universe. The earth is flat, and the sun travels over it by day and under it by night.

Night is dark and the time for rest and sleep. At least, this is so for most people, but long before men settled down to build cities and live in large communities, their social organization was in small nomadic tribal units. Their food was not obtained from farms, however small and primitive, but gathered from the surroundings in which the tribe found itself. Fruit was picked, root plants dug up, and animals hunted. Some hunting was carried out at night; the hunters had to know something of the night sky because on some nights there was a bright shining body in the darkness, and sometimes there was not. This body was not so bright as the sun, but it could give considerable light, enough to make night hunting possible.

The moon, as we call this body, rose in the east and set in the west just like the sun and it seemed obvious that it too traveled back beneath the earth. Its power to shine so brightly at night also made it an obvious choice as another god, to be worshipped and offered sacrifice. But the moon was a complex body for, as the days passed, not only did it rise and set, but it changed its shape. Sometimes it was a crescent, sometimes a disk, sometimes a shape in between the two. Surely a god that could behave in this way must be of great magical importance, and doubtless this was one reason why it entered into many rites and customs. Moreover, the fact that the cycle of changes in the moon's shape kept pace with the cycle of bodily changes in adult women gave it great significance. Why it changes shape is not known at this stage of development—perhaps it is just part of its magical power.

From a practical point of view, the moon was important not only for the light it gave to the hunter, but also because its changes were comparatively rapid: the change in shape—its phase—could be observed from night to night. For people to whom writing was unknown, counting was confined to reasonably small numbers, and the changing moon offered a simple and convenient way of noting the passage of time. To use the alteration between night and day was all very well, but it was only too easy to reach large numbers and become confused; the moon and its phases gave a useful guide. Thus, the earliest calendars were lunar, based on the moon's phases, and became an essential factor in dating religious festivals.

The night sky displayed a host of smaller lights—tiny pinpoints that we call stars. Always forming the same patterns in the sky, they seemed to be permanently fixed, the moon moving across them as the weeks passed. What were these lights and to what were they attached? Probably a detailed explanation of their nature was beyond any primitive man—it would be sufficient to give them a name and think of them as celestial torches lying a great distance away—but the problem of where they were fixed did not seem so difficult. The reason for this lies in the appearance of the night sky. If we go out into the country, away from city lights, and look at the stars and sky, we shall find it difficult, even today, to think of the sky as anything but a gigantic dome stretching over our heads. Only by remembering that the astronomer now knows that the stars are spread out in space and at different distances from us can we really tear our minds away from a dome-shaped heavens. If left to ourselves, without any previous facts or ideas to guide us, a dome is the one answer to the question of how the stars remain together in the same patterns and why those directly above us stay there and do not fall down on top of us. So the earliest astronomers constructed a universe with a flat earth and a dome for the sky.

About 5,000 years ago, during the neolithic, or new stone, age, when mankind began to use polished stone implements, there arose the idea of cultivating plants. This was probably a consequence of discarded nuts, fruit pips, and grass seeds sprouting near the camp sites of prehistoric man. And once cultivation began, the wandering tribe had to settle down to tend its crops and to domesticate such animals as the ox, pig, sheep, and goat, which it had previously hunted. All this affected tribal astronomy because, once settled in an agricultural community, the seasons became the predominant factor in the cycle of time. No longer was the moon the guide because the men hunted less and spent their time sowing and harvesting crops and tending their livestock. The

monthly cycle of the moon's phases gave way before the longer
cycle of the year, which was determined by the motion of the sun
in the sky.

How, it is worth inquiring, was it possible to observe the
movement of the sun against the background of stars? The moon
presents no problems because the stars can always be seen when it
is visible, and its path across the heavens can be observed quite
easily. But it is quite different as far as the sun is concerned, for
when the sun has risen, the stars are quite invisible. How could
primitive man be sure that the stars were still in the sky when the
sun was shining?

There are three ways in which he could obtain evidence, and
probably he was aware of all of them. In the first place there was
the rare occurrence of observing the moon passing directly across
the sun's disk and completely eclipsing or blotting it out. Such
total solar eclipses are not rare in themselves—on the average,
there are eight every ten years, but they are seldom all seen by
people living in a particular area. For instance, between the years
1970 A.D. and 2000 A.D. there will be twenty-three total solar
eclipses, yet only two of these can be observed in the United
States, each from a narrow path not eighty miles across, one in
Florida and the other just touching the states of Washington,
Idaho, and Montana. Besides being a rare sight at any one place,
even when total eclipses can be seen they last for no more than a
few minutes, because the moon's shadow moves quickly across the
earth: for example, of the twenty-three eclipses mentioned, the
longest takes only slightly more than seven minutes, but most take
fewer than five. However, when a total solar eclipse does occur,
even if it lasts no more than a minute, that minute is so breath-
taking that it must remain in the memory forever.

We may expect primitive man to have observed total solar
eclipses and to have been overawed by what he saw—the sudden
extinction of the sun's light and heat when the moon covers the
solar disk, the appearance of a pearly light around the sun, the
twilight that descends all around, and the fact that animals go to
rest. But from the point of view of the motion of the sun across the
sky, the most significant thing—and one of the most impressive—
is the sudden appearance of stars in the darkened sky. A total

eclipse is so memorable that it would have been talked over time and again and become part of folk legend, and the appearance of the stars would therefore have become well established. Man would know that the stars and the sun can be in the sky at the same time.

The second line of evidence that the sun and stars are in the sky together comes from the simple everyday observation that just before the sun rises some of the brighter stars that are overhead or close to the opposite horizon remain visible for a short time. Only as the sun rises above the horizon do the bright stars vanish in the full glare of daylight. The opposite effect can be observed at night, the brighter stars becoming visible before all the sun's light has completely gone. And this kind of observation would be particularly noticeable when the bright morning star or evening star was to be seen (that is, what we now know to be, respectively, the planets Venus and Mercury).

Third, there was a more sophisticated observation that later evidence, to be considered shortly, shows was certainly made as settled communities began to grow and some men, at least, had plenty of time to spend contemplating the heavens. These men would, perhaps, be those who tended flocks out in the open and, more especially, the priests, because celestial observation helped them fix the dates of religious festivals. This sophisticated observation depended on observing the patterns, or constellations, of stars. During the whole cycle of a year, all the various constellations would be seen and gradually, as men became used to them, they arranged them into definite patterns, depicting animals and legendary characters. Once the patterns were recognized, it was noticed that just before sunrise a particular constellation could be seen above the eastern horizon at one time of year, a different constellation at another time of year, and so on. Again, the same kind of observation could be made on the western horizon, just after sunset, but the constellation would be a different one. This showed that the constellations remained the same, year in and year out and that the presence of the sun did not destroy them.

This sunrise-sunset constellation observation made evident another important fact—the motion of the sun among the stars. Continued observations by the priesthood of early communities had led, not only to a separation of the stars into constellations, but also to the recognition of the fact that the sun and moon both traveled among them along similar paths. Along these paths lay a collection of constellations depicted mainly by animals, which have become known as the zodiac, or band of living figures. Their order was also recognized; thus, using our present-day names, it begins with Aries (The Ram); then comes Taurus (The Bull), Gemini (The Twins), Cancer (The Crab), Leo (The Lion), and so on. Now if, say, Aries was observed on the western horizon just after sunset, and Gemini on the eastern horizon just before sunrise, then it is clear that the sun must lie in between the two constellations. A glance at the list shows that it must therefore lie in Taurus, because Taurus is between Aries and Gemini.

This technique of noting the stars near sunrise and sunset has become known as observing the heliacal risings and settings, from the Greek word *helios* meaning "sun," though it was practiced long before Greek times by the Egyptians and the people of Babylonia who lived in what is now Iraq, northwest of the Persian Gulf. In the latitudes of these countries (30° to 33°), the sun rises and sets at a fairly steep angle in the sky and twilight is short; heliacal risings and settings are an admirable way of determining the seasons because the stars can be seen immediately after sunset and until a few moments before sunrise. But this is not so in countries farther from the equator, where the sun rises and sets at a more sloping angle to the horizon and there is a long period of twilight in the evening and at dawn. In consequence, heliacal risings and settings are less useful, because the sun has to be far below the horizon before the stars close to the horizon become visible. In these countries mankind was forced to develop a different technique, whereby the sun and not the stars were observed. This was possible because at different seasons of the year the sun rises and sets at different points on the horizon: in northern midwinter it is at its most northerly point, and at midsummer at its most southerly. About 2000 B.C. this fact was well realized, and in some countries huge stone circles were built so that the priest-astronomer could stand at a particular point within the circle and watch near which stone the sun rose and set. The circles measured up to ninety-five yards or more in diameter, and the stones themselves were frequently immense; at Stonehenge near Salisbury, England, where part of a very large and complex series of stone circles still remains, some of the stones weigh more than thirty tons. Such massive constructions are called megaliths from the Greek *megos* meaning "huge," and their builders are known as megalithic men.

Megalithic structures were widespread. In Britain, where most research on stone circles has been carried out, more than 450 have been discovered, but in western Europe a number of similar megaliths are known. In Egypt and in Babylonia, huge structures were also built at least as early as 3000 B.C. The most notable of these, constructed from about 2400 B.C. until at least 1,000 years later, are the pyramids, colonnaded temples, and ziggurats, or stepped towers. The pyramids were tombs and the ziggurats were temples and both were usually laid out with the sides aligned in north-south and east-west directions. For instance, in the great colonnaded temples to the sun god built at Karnak and Abusir in Egypt, the giant columns stretched in an east-west direction so that the sun would shine directly down between the lines twice a year.

The astronomical alignment of burial places—the pyramids—and of ziggurats and colonnaded temples to the points of the rising and setting of the sun underlies two facts: the importance of the sun and other celestial bodies in early worship, and the astronomical duties of the priests, for the priesthood contained most of the learned members of the community and was responsible for preparing a practical calendar. The development of a workable calendar was vital. It was difficult because it had to incorporate two quite different methods of reckoning time—one based on the moon and used for determining the dates of religious festivals, another based on the sun, which determined the dates for such agricultural activities as sowing and reaping. The difficulty in fitting the two together was mathematical. The moon takes twenty-nine and a half days (more precisely 29.53059 days) to complete its cycle of phases from, say, full moon back to full moon again—the period of a synodical month—but the earth completes one orbit of the sun in 365.2422 days. This last period is the time taken for the sun to move from one position in the sky to the same apparent position—from the moment when the hours of day and night are equal in the spring (the vernal equinox) back again to the next vernal equinox. Owing, as we now know, to the complex motion of the earth in space, the period from vernal equinox to vernal equinox (the tropical year) is not the same as the year measured with reference to the motion of the sun among the stars. The star motion, or sidereal, year is a little longer, namely, 365.25636 days. Whether we count the days in a sidereal or tropical year, we find that an exact number of synodical months cannot be fitted into the period. This is also true even if we calculate the month differently—by the time the moon takes to complete an orbit with reference to the stars instead of from full moon to full moon. Thus, the problem facing those who needed to fit a lunar calendar based on the moon's motion to a solar calendar based on the apparent motion of the sun was a very difficult one to solve.

The astronomer-priests of Babylonia attempted to reach a solution by the most complex methods, using a lunar calendar as a basis for their calculations. Days were inserted in the calendar from time to time when it was found necessary to move closer into step with the seasons. There seems to have been little regularity in deciding when these intercalary days should occur until sometime around 300 B.C., though nearly a century earlier the Babylonians had discovered that every nineteen years contains 235 lunar months, a fact that would allow them to compute a rule for adding days to the lunar calendar to bring it into line with the solar one. This nineteen-year cycle, which is correct to within two-tenths of a day, was also discovered by Meton of Athens in Greece some fifty years earlier and is usually known as the Metonic cycle; possibly it was Meton's discovery that the Babylonians adopted.

In Egypt the agricultural year began in early July with the annual flooding by the Nile of the low ground either side of it. By September every village stood like an island with about six feet of water covering the surrounding land, and then the waters subsided. The winter was mild, and wheat readily germinated so that the harvest was gathered by mid-April. This was a regular cycle and very early led the Egyptians to recognize a year of 365 days, thus forming a civil solar calendar in the early years of their civilization—probably by 3000 B.C., if not before. This calendar was in error by one-fourth of a day, but adjustments could be made, and were made, when the calendar became too far out of step with the seasonal changes of the Nile. The addition of an extra day occurred at the end of the year because the Egyptians used twelve regular months of thirty days each and added a period of five days (or more) at the end of the last month. Each day lasted for twelve hours, the hours having different lengths according to the seasons of the year because the twelve hours had to be fitted in between sunrise and sunset. It was an excellent practical scheme of great simplicity.

The solar calendar was later determined by the heliacal rising of the very bright star Sirius, known to the Egyptians as Sothis, the Dog Star. Sirius appeared at dawn at the beginning of the flooding of the Nile, and, in due course, from observations of the date of the heliacal rising of Sirius they realized that the year consisted of 365.25 days. This led them to compute the Sothic cycle, a period of 4×365 days (1,460 days), after which everything once again occurs on the same date.

The observations of Sirius and of other heliacal risings and settings make it obvious that the Egyptians knew the constellations well, but the way in which they plotted them out was not very satisfactory. They were familiar enough with the brighter stars, but they divided the sky itself into decans, each decan containing those stars that rose heliacally during a period of ten days. This system was of little use astronomically, because the boundaries were vague; indeed, today archaeologists are still confused about where one decan ended and another began, even though they are men of intelligence and there are many low-relief carvings for them to examine. The Babylonians were more astronomically minded than the Egyptians even though their calendar was neither so simple nor so effective. Even as far back as 2700 B.C., the Sumerians, the first civilized people to occupy the fertile land near the mouth of the Euphrates River, began to develop mathematics and to apply their knowledge to their observations of the sky.

During the fifth century B.C. astronomical observations, particularly of the sun, were made in India, but the Indians' knowledge seems to have been gleaned from the Babylonian civilization with which they had trade connections by an overland route through what are now Iran and Pakistan. Far to the east in China, a vast collection of astronomical knowledge was being amassed at the same time, quite independently. Seven hundred years later, however, some of the Babylonian ideas and particularly their pattern of the zodiac, arrived in China and were adopted there, but before this the Chinese developed their own system of constellations and their own calendar. Between 1500 and 1000 B.C. they used a lunar calendar composed the three groups of ten days to give a month, and a longer cycle of sixty days. They observed the stars and took special note of the moments when the bright star that we call Antares in the contellation of Scorpio (The Scorpion) was due south at dawn or dusk, relating this period of the year to their lunar calendar. They also observed the constellation Orion and the handle of the Dipper.

During the next 600 years more star formations were noted, and they adopted a pattern of twenty-eight mansions, or houses, into which they divided the sky in order to determine the positions of the sun and moon more accurately. Spring was determined by the southing of Alphard, the brightest star in the constellation Hydra, and solar eclipses were observed in detail. It was believed that eclipses were caused by a dragon swallowing the sun and that light could only be restored by frightening the dragon into disgorging its prey. There is, in fact, a well-known story that two astronomers, Hsi and Ho, who were charged with the duty of preventing eclipses, once failed owing to negligence (being more interested in living a comfortable life than in their work) and were subsequently executed. The story, however, is a distorted legend, there never were two such astronomers, the names referring to an imaginary being Hsi-Ho who in some legends is the mother of the sun and, in others, the charioteer who carried the sun across the sky.

We see, then, that beginning with a curiosity about the sun, moon, and stars and forced by the need to determine a calendar based on the seasons, man began to study the heavens. In every civilization, the skies attracted the attention of thinking men. So far we have been concerned mainly with the practical use of the knowledge that was gradually collected. But what were the earliest ideas about the whole universe? How powerful were the sun, moon, and stars? The sun was worshipped as a god, and this was not unnatural considering its obvious power to assist the growth of crops. The moon also was worshipped, for it was a somewhat mysterious body as well as a convenient timekeeper. The fixed stars appeared in patterns that could be recognized, patterns that were like animals and people of legend and so became linked with qualities that were a familiar part of life. Thus it seemed quite natural to expect that the seemingly everlasting celestial bodies influenced mankind.

The belief in the power of celestial bodies to shape events on earth was heightened by the fact that the earth itself was thought to lie in the center of the universe, with the stars, sun, and moon moving on a dome overhead. No one knew how large the dome might be—clearly it was beyond the reach of man and higher than the tallest mountain—yet it appeared close enough to exert great power. It was thought to be the source of rain, clouds, and storms and was therefore intimately connected with everyday life. The so-called science of the power of the heavens on mankind—astrology—seemed to be an obvious part of astronomy and, in fact, was one of the reasons for the great interest taken in the heavens and all that occurred there. No separation between astrology or astronomy could then be made, though the various civilizations often employed different methods in studying the heavens.

Even though a dome of heaven was accepted the world over, descriptions of it varied widely. To the Egyptians, who worshipped a host of gods and goddesses, the sky was the body of the goddess Nut (see Figure 1–1) and the earth, the body of the god Qeb with the god of the air above. The sun and moon were deities that traveled in ships across the sky. The Polynesians and the Mexicans also thought of the celestial bodies as gods and goddesses. The Chinese were rather more scientific in their approach for they considered the heavens similar to a bowl turned upside down, with the earth itself like a smaller inverted bowl lying beneath. This Chinese view of the universe, however, was much later than the Egyptian—probably around the sixth century B.C. compared with 2000 B.C.—and as with much of their early astronomy seems to have been derived from the ideas current in Babylonia.

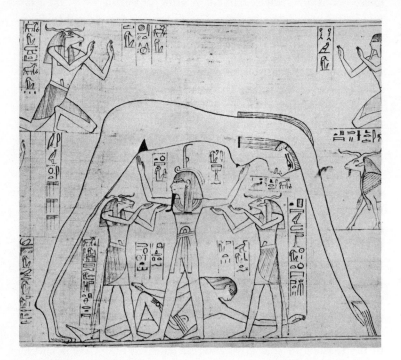

FIGURE 1–1. The Egyptian picture of the universe, showing the goddess Nut, whose body represented the starry sky, the reclining figure of the god Qeb as the flat earth, and the god of air, Shu, standing up and supporting Nut; on either side of Shu are two spirit gods. From the papyrus of Nesitanebtâshru, circa 970 B.C., presented to the British Museum, London, by Mrs. Mary Greenfield. Courtesy of the Ronan Picture Library.

Here, while the Egyptians were holding to their ideas of a divinely populated universe, a more truly material picture was being drawn. The dome of the heavens was a real solid dome, with holes through which the waters above the sky could rain down on the earth; it was supported along its rim by a ring of mountains separated from the flat earth by a vast encircling sea. This was indeed a scientific picture, even if a very primitive one.

The belief that the heavens were a dome persisted for an extremely long time; it was not until about 700 B.C. that a new idea began to emerge. It arose in the Ionian civilization of the eastern shores of the Aegean Sea (what is now the west coast of Turkey) and its many offshore islands. With some astronomical knowledge derived from an earlier civilization that had flourished under King Minos at Knossos on the island of Crete, the Ionians developed an independent view of the universe. From the Minoan civilization they derived a knowledge of the constellations that was very similar to that we use today, but they seem soon to have rejected the idea of a dome of heaven. From the epic poems of Homer and of Hesiod we find that though the earth was still thought to be flat, they conceived of the heavens as part of a giant hollow globe. Why they did so we cannot be sure, even though it is clear that it cannot have been owing to the knowledge we now possess from seeing stars wherever we travel over the world (which we know to be a globe). The probable answer is that the Ionians, like the Greek civilization into which they were to merge, had a passionate love for beautiful forms, and chose a globe-shaped universe simply because a sphere is a more elegant and balanced shape than a dome or hemisphere. But of one thing we can be certain, this new picture of the universe became the important view of the Western world, completely replacing the older dome of heaven; in China it seems to have appeared about three centuries later. Historically, the great significance of the spherical universe is that it remained unchallenged for more than 2,000 years.

2 ✳ The Planets and Their Motions

What we observe in the sky depends on how carefully we look. It is surprising what can be noticed if only we take the trouble, and if we observe regularly; this, of course, is what the earliest astronomers did. Shepherds, keeping an eye on their flocks, would constantly be gazing at the sky and become familiar with it—after all, they would have little else to do most of the time. Other careful observers would be the priests and the scholars and philosophers; the first because they wanted information to help them in constructing and using their calendars, and the others because they were curious. Sometimes, the priests and philosophers were the same men, but no matter. What they observed was an immense pageant of moving constellations, awe-inspiring in its vastness, its beauty, and its regularity. Only one thing marred the perfection of the great system of the heavens—the irregular behavior of five stars.

The irregularities were a nuisance. They did not fit into a nice simple scheme of the universe where each star kept its appointed place on the celestial dome. Instead they wandered about, sometimes being in this constellation, sometimes in that, and on occasions were not visible at all. And because they were bright, one could not ignore them. Somehow their motions had to be accounted for as part of the grand scheme of the cosmos. But as the observers watched regularly, week after week and month after month, they found that the problem was even more complex than they had at first supposed. There seemed to be two different kinds of wandering star, one that could be seen only near sunset or near dawn, the other that could be observed at any time of night.

Of the sunset-sunrise stars—the morning or evening stars as they came to be called—it was gradually realized that there were

two separate bodies, each of which was sometimes seen in the eve-
nings and at other times in the early morning. As far as the night
wanderers were concerned, three were recognized. The names
given varied with the civilization observing them and the language it used—Nir-dar-anna, Hesperos, Aphroditēs astēr for what
we now call Venus and Apollōn or Hermēs for our present-
day Mercury. The night wanderers also had various names; Mars,
Jupiter, and Saturn are the ones with which we are familiar. And
for convenience, it will be as well if we stop using the word
"wanderer" and substitute "planet," which is derived from the
Greek *planasthai* meaning "to wander."

Whatever the name, what the astronomer wanted was to account for the motions—to explain why Mercury and Venus were
only seen near sunrise and sunset and why the other three planets
could be seen at any time of the night (but not every night). He
had also to explain why Mars, Jupiter, and Saturn went in one
direction among the stars, stopped, and then moved backward for
a time—in fact to account for why they performed loops in the
sky. Of all the early civilizations the Egyptians alone seem to have
taken virtually no notice of the planets, and they certainly worked
out no theory to account for them nor made any close examination
of their movements. But this was very far from the case with the
Babylonians.

Two factors made the Babylonian contribution to planetary
studies important—their observations and their mathematics. Far
more advanced than any other civilization in the study of arith-
metic and algebra their mathematics enabled them to specify not
only where the planets were to be seen at any particular time, but
also how they moved from day to day. Using a method of counting that was a mixture of a scale of ten (such as we use) and a
scale of sixty (which we still adopt when we measure time and
angles), by 2000 B.C. the Sumerians were able to do many
mathematical calculations that others could not do until more than
1,800 years later.

Yet the upheavals that followed three conquests of the country,
and the fact that in those days developments took place more
slowly than they do now, meant that it was a very long time before
this mathematical knowledge was applied to the problems of de-

scribing the motions of the moon and of the planets that had been observed for at least 2,000 years. All the same, in the last three centuries B.C., Sumerian mathematics was used by the Babylonians for solar, lunar, and planetary studies, with the result that they drew up tables of numbers representing the behavior of the planets.

They had already made tables of the moon's motion and were particularly interested in those moments when it would appear on the eastern horizon and would disappear on the western, and so they became interested in these same moments for the planets. Their tables were sufficiently complete and their mathematics advanced enough to enable them to work out the places in the sky at which the planets would be at any given date, though not with the accuracy that was to come during the second century A.D.

The Babylonian study of planetary motions and their mathematical analysis of them did not, as far as we know, lead them to any complete planetary theory. All we can say is that they must have assumed that the earth was fixed in space and that the planets moved round it in some way that gave rise to the loops in their paths. It is to the astronomers and philosophers of ancient Greece that we must turn for a theory of planetary motion, because the Greeks took a rather different attitude than the Babylonians'. The Babylonians first made observations, next devised mathematics to describe what happened, and only then started to think about how this could be theoretically explained. Considering that their observations were incomplete and that planetary motions are extremely complex viewed from the earth, it is hardly surprising that they could not reach a definite conclusion. The Greek approach was almost completely the opposite—having recognized that planets move, they then constructed a theory; only after this did they observe the motions in detail and adjust the theory to correspond. But, as we shall soon see, how much they could alter their theory was limited by some of their basic beliefs.

The earliest Greek philosophers had various ideas about the planets, when they considered them at all. Some thought they were nearer than the sun but farther than the moon. The first man of real importance in planetary theory was Pythagoras, who lived between 580 and 500 B.C. Usually known now for his theorem about the sides of a right-angled triangle, he was a preacher and mystic, the founder of a religious order whose members lived a simple life, and a teacher of mathematics and ideas about the natural universe. Pythagoras seems to have written nothing, but his teaching was passed on by followers; from them we learn that he thought of the earth as a round globe, not a flat disk as most people then believed, that he realized that the morning and evening stars could be the same body appearing in different positions, and that he decided that the motions of the sun, moon, and planets could best be described by assuming that they moved in circles.

So far Pythagoras' planetary teaching is not outstanding, but he believed strongly in a divine harmony in the universe; as far as we know, he was the first to speak of a music resulting from the motions of celestial bodies. His celestial harmony is important because his followers believed in it too, and coupled to the idea that the planets moved in circles, they used it to lay the foundations of all later theory. Stated simply, they taught that the motions of the planets could be explained by regular motion in a circle. In other words, the planets circled the earth and did so at a set speed that was unchangeable. Only this kind of motion was perfect enough to be suitable for celestial bodies—not an argument that would find favor with the scientist today, but one that seemed appropriate enough 2,500 years ago, when science, art, and religion were not separated as they are now. Philolaos, one of Pythagoras' followers, went so far as to suggest that the earth also moved in a circular path. However, he thought that sun, moon, planets, and earth moved in circles around a fire fixed in the center of the universe, and not around the sun as some people later mistakenly believed he had taught. We do not see the central fire because the part of the earth on which we live is always turned away from it. Though Philolaos did not think that the earth rotated on its axis, he believed it took only twenty-four hours to complete one circuit of the central fire, and so provided an explanation of how the stars and other celestial bodies rise and set, without having to suppose that the whole universe rotated. But his main aim in having a moving earth was to introduce a counter-earth that circled the central fire at the same speed as the earth but in a closer orbit. His purpose in having a counterearth was that, with the earth, the sun, the moon, the five planets, and the sphere of the fixed stars he had a universe that contained ten bodies, and this was aesthetically and religiously satisfying, because ten was a number of immense significance in Pythagorean mystical teaching.

Philolaos' scheme of planetary motion had little effect on later thought, and most held to the idea that the earth was immovable and fixed firmly at the center of the universe and that the planets moved at a constant speed in circles about it. This basic scheme of Pythagoras was adopted by the great philosopher Plato, who was a pupil of the equally famous Socrates. After traveling widely and being kidnapped by pirates, Plato settled in Athens in 387 B.C., and began teaching at the Academy, which was on the outskirts of the city and set in a park sacred to the hero Acadēmos. Plato not only adopted the explanation of Pythagoras about planetary motion, but also developed some of the mathematical mysticism of his predecessor. He placed the celestial bodies in an order of distance from the earth, with the moon the nearest, the sun next, followed by Venus, Mercury, Mars, Jupiter, and Saturn —an improvement over earlier ideas. But by and large Pythagoras' and Plato's ideas of planetary motion were most unsatisfactory. The sun's motion could be explained by saying it moved round the earth in a circle at a constant speed, provided one did not observe its behavior in too much detail, and the same was true, but to a lesser extent, for the moon. For the planets it was totally inadequate. It fell to Eudoxos to provide a better scheme.

Eudoxos was born in 408 B.C. in the seaport of Knidos on the west coast of what is now Turkey. After studying under Plato, he traveled abroad and finally settled back in Knidos. A great mathematician, he excelled in geometry—the particular branch of mathematical knowledge for which the Greeks had a flair. Why they were so adept at it we do not know, but it is clear from looking at their works of art and taking into account such ideas as the spherical universe which, as we have seen, they preferred to one composed of only a hemisphere, that they possessed this great feeling for shape. Their artistic interest in the proportions of the human body was another example, even though their choice of vital statistics was not the same as we adopt today! It was from a study of shape that Eudoxos was led to his ingenious planetary theory. Known as the theory of homocentric spheres, it was a brilliant breakthrough into more precise astronomy, because it provided a basis for explaining planetary motion by using regular motion in a circle and keeping to the belief that the earth is fixed at the center of the universe. In essence, what Eudoxos did was to consider a great number of spheres, one lying inside the other (homocentric) like a whole series of balls, each smaller than the one outside it. The center of all the spheres lay exactly in the center of the earth. The breakthrough that Eudoxos achieved was that he used more than one sphere for each celestial body, and placed the axes about which they spun in various directions. This probably sounds rather complex—and his whole system was complex to some extent—but we can readily grasp what he achieved if we consider, say, the four spheres he used to account for Jupiter's motion. Here what he had to do was to explain (1) how the planet rose and set each twenty-four hours, (2) how it completed its circuit among the stars every thirteen months, and (3) how it performed its loops in the sky. Eudoxos achieved his aim (see Figure 2–1) by using four spheres. The first, A, spins round once every twenty-four hours on an axis that lies through the earth's north and south poles. The second sphere, B, spins round once every twelve years on an axis that lies tilted 23° from the north and south poles—thus taking into consideration the fact that the path of the planet lies close to the path of the sun—the ecliptic. To cover the loops in the sky, Eudoxos had to use two other spheres, C and D. Sphere C rotated on the same axis as sphere B, but moved round once every thirteen months (the period the planet takes to make a circuit of the sky counted from the sun back to the sun again). Sphere D also rotates at this speed but in the opposite direction, and its axis is inclined at 90° to the axis of sphere C. The planet was supposed to be fixed to the equator of the fourth sphere, D.

Every planet possessed its four spheres, though only three were required for the sun and three for the moon. Even so the system did not explain how eclipses happened, nor the apparent changes in size of the sun and moon, and though it was reasonably satisfactory for Mercury, Jupiter, and Saturn, it failed to some extent for Venus and more so for Mars. Yet, with all its faults, it represented a great stride forward in the explanation of the cause of planetary motion and was adopted by Aristotle.

Aristotle, born in 384 B.C. at the seaport of Stagira, on the north coast of the Aegean Sea, was, like Eudoxos, a pupil of Plato. He became the tutor of the future Alexander the Great and later,

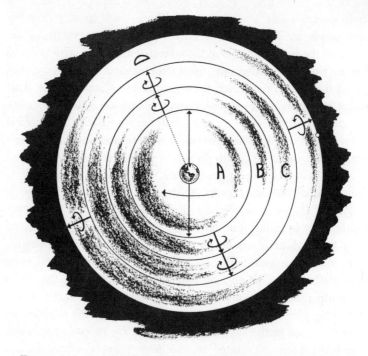

FIGURE 2–1. The spheres of Eudoxos for explaining planetary motion. The complicated nature of the system may be judged from the fact that spheres for only one of the seven planets are shown here.

in 337 B.C., established his own academy in Athens. It was called the Lyceum and much of the work was done while Aristotle and his pupils walked in the shady groves of the school, a habit that led to their being called peripatetics (that is, walkers). Aristotle concerned himself with every branch of philosophy—he discussed morality and ethics, political matters, economics, and natural science. His observations of animals and his general biological knowledge were highly advanced, but it was his teaching about physics and astronomy that was later to have the most far-reaching effect.

Aristotle adopted the homocentric sphere theory of Eudoxos, but modified it slightly so that the number of spheres to account for all the moving bodies rose from twenty-seven to fifty-five. He differed from Plato in the order he adopted for the planets, placing Mercury nearer the earth than Venus. The outermost sphere of the universe, which contains the stars, was, he thought, turned by God, and he believed it was from this motion that the rest of the spheres derived theirs. The earth, he argued, was motionless, but it is worth noting that he did consider the question of its possible motion in space. He rejected the idea because there were too many arguments against it: (1) The earth is too heavy to move. (2) There is no evidence of its motion. (3) He appreciated that if the earth did move then at different times it would be closer to different stars (because all the stars are fixed on a sphere), and this should lead to observational differences that no one had ever noticed. We now know that these arguments were mistaken because of incorrect views about motion and about how close the stars are to the earth, but in the fourth century B.C. they seemed both sound and logical. It is only fair to state that Aristotle's arguments to prove that the earth is not flat but spherical are still as sound as they ever were. He mentioned how curved the shape of the earth's shadow is when we see it move across the face of the moon during a lunar eclipse: he quoted travelers who mentioned that as one travels north or south different stars become visible over the northern or southern horizons; finally, he pointed out how, wherever one may be, things always fall straight down to the ground and never sideways. This final argument was based on Aristotle's belief that everything falls to the center of the earth, though we must not read history backward and imagine that he had an idea of gravity. For Aristotle, solid things fell toward the center of the earth because this was their natural place—it was their nature to do so—whereas such liquid things as water spread around the earth (as we can see if we spill any) because it is the nature of liquid things to spread over the earth's surface. Aristotle adopted four basic chemical elements—earth, air, fire, and water—but he believed all the celestial bodies to be of a quite different material. This, he argued, was a necessary conclusion because everything on earth suffered change, whereas all evidence pointed to the heavens being eternal. He therefore required a fifth element to account for this.

Today Aristotle's views may seem strange because we are brought up with quite different thinking about what elements are, and we do not believe things have one natural place in the universe. But when Aristotle looked at the universe and tried to form what he saw into one grand comprehensive scheme, he welded his belief in four elements and his teaching about natural properties and places into one vast plan that embraced the planets, the stars, and even divine providence. Though there is no chance of our now accepting Aristotle's universe, we must not fail to appreciate what a truly magnificent achievement it was.

But even in Aristotle's time things did not stand still, and there were soon other ideas for the philosopher to consider. About 340 B.C., while Aristotle was still alive, Hērakleidēs (born in the district of Pontus near the Black Sea and not to be confused with another Hērakleidēs born in Tarentum in the southeast of Italy) proposed the doctrine of a rotating earth. Brought up on Pythagorean ideas, and yet familiar with Aristotle's system, Hērakleidēs felt that to assume the immense sphere of the stars rotated once every twenty-four hours was going too far. To imagine that the much smaller earth spun on its axis was more in keeping with a sense of proportion. Hērakleidēs also thought that the behavior of Mercury and Venus suggested that they orbited the sun rather than the earth. However he did not go so far as to adopt a truly heliocentric or sun-centered system for all the planets: this was left for Aristarchos.

Aristarchos was educated at the Lyceum by one of Aristotle's successors, and his main work was done sometime about 280 B.C. He attempted to measure the distances of the sun and moon, and though observing techniques were not accurate enough to allow him to work out how many miles they are away, he was able to conclude that the sun is between eighteen and twenty times farther away than the moon. We now know this is wildly wrong—the sun is 397 times farther away—but it was at least a step in the right direction, and far nearer the truth than were the views of two centuries earlier, which supposed the sun to be only just up in the air, a little way above the earth. But, Aristarchos' main claim to fame is that he suggested that all the planets, not only Mercury and Venus, orbit the sun; he included the earth as a planetary body. He realized that such a view meant that the sphere of the stars must be an immense distance away (otherwise we should see changes as the earth moved in its orbit). However, as far as we know, Aristarchos wrote nothing about his theory, and it seems soon to have been cast aside. This is not surprising when we remember that if the sun was to be the center of the universe, the wonderful scheme of physics and the heavens worked out by Aristotle was to be upset. Aristarchos was a mathematician, and other philosophers doubtless treated his sun-centered theory as an interesting mathematical exercise, a view he may even have taken himself. It would allow one to work at some of the facts of planetary motion more easily than using a geocentric or earth-centered universe, but it was not worth upsetting the whole of the physics of moving bodies and of the elements to gain the advantage of a slightly simpler geometry. And what would be the divine purpose of creating a universe where there was an incredibly large gap between the last planetary sphere (of Saturn) and the sphere of the stars. No, from the practical point of view, it seemed nonsense.

There was every reason to keep to a geocentric universe, but if the earth were fixed in the center, certainly there was a need to improve on Eudoxos' theory of homocentric spheres. What could be done? The answer came from another mathematician, Apollonios, who was born in Perga (now in the south of Turkey) about 262 B.C. Apollonios was brilliant at geometry and investigated the properties of such curves as the oval or ellipse, the parabola, and the hyperbola, all of which can be obtained by cutting slices from a cone and are known as conic sections. But his great contribution to astronomy came with his use (and probably invention) of the epicycle and deferent to explain planetary motion (see Figure 2–2). Essentially this was a simple geometrical way of accounting for the loops as well as the orbits of the planets, taking the earth as center. The large circle, with its center at the center of the earth (Figure 2–2), is called the deferent, or carrying circle, because it carries on its rim the center of another and smaller circle, the epicycle, or rolling around circle. The planet was thought of as being fixed to the rim of the epicycle, which rotated as its center was carried round the rim of the

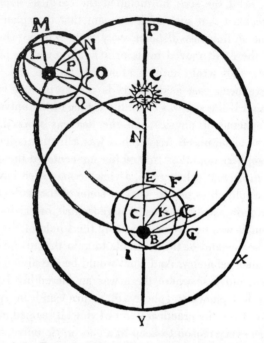

FIGURE 2–2. Epicycle and deferent used for explaining planetary motion. The epicycle—here double to give a more precise account of observations, and shown in two positions—is on the upper left and lower middle. It carries around the planet, shown in this case as a crescent to represent the moon. The dark blob in the center of the epicycle moves around the circumference of the large circle on which it lies, and whose center is at N. This drawing, showing the earth at the center of the moon's epicycle and also in orbit about the sun, illustrates Copernicus' ideas (some 1,800 years after the time of Appollonios). This shows how important the epicycle and deferent were and why they continued to be used for so long. From A'. Kircher, *Magnes: sive arte magnetica* (Rome, 1643). Courtesy of the Ronan Picture Library.

deferent. The number of loops a planet would perform depended only on how many times the epicycle spun around each time it moved once around the deferent.

Apollonios' scheme was put to great use by two other Greek

astronomers whose work really completed early astronomy, because they managed to account for all planetary motions with sufficient accuracy to agree with the observations of the time. These were Hipparchos and Ptolemy. Hipparchos was born about 190 B.C. and was a great observer. Most of his work was done on the island of Rhodes; the majority of it has not survived and our only knowledge of his astronomy comes through Ptolemy. Hipparchos adopted slightly different schemes from the deferent and epicycle; they were called the fixed eccentric and the movable eccentric—both had been invented by Apollonios. The fixed eccentric was a circle whose center lay not at the center of the earth, but a small way off. As a body moved at a regular speed round the center of the circle, it appeared to move at an irregular pace when viewed from the earth. Hipparchos used the scheme to explain the motion of the sun. For the moon he used the movable eccentric, which was more complicated because it had two circles —a small one centered on the earth and a larger one whose center moved round the rim of the smaller. It was rather like the deferent and epicycle worked in reverse, and it gave the same results. But in spite of this, Hipparchos was unable to account accurately for the observed results of planetary motion—perhaps because he was the greatest of all the Greek observers and put observational results before philosophical theory. At all events, he contented himself with classifying his planetary observations, and it fell to Ptolemy to produce a really powerful mathematical analysis of their motions.

Ptolemy, whose real name was Claudius, was born in the city of Ptolemais and carried out his epoch-making research sometime between 120 and 150 A.D. He worked at the Library and Museum at Alexandria, a research institution of immense reputation and broad outlook (an ancient parallel of today's Princeton Institute for Advanced Study). Here the famous geometers Apollonios and Euclid had studied and taught, and one of its directors was Eratosthenēs, who measured the size of the earth correct to within fifty miles. The city of Alexandria had been founded in 332 B.C. by Alexander the Great, who had just conquered Egypt, and when his immense kingdom was broken up after his death in 323 B.C., Ptolemy Soter, one of his generals, gained control of Egypt and established the Library and Museum, as well as his own ruling dynasty. In medieval times many scholars became confused be-

FIGURE 2–3. Ptolemy using a quadrant and attended by the goddess of astronomy. An armillary sphere is shown lower left. As usual in medieval times, Ptolemy is incorrectly shown wearing a crown. From G. Reisch, *Margarita Philosophica* (Heidelberg, 1508). Courtesy of the Ronan Picture Library and the Royal Astronomical Society.

tween Ptolemy the astronomer and Ptolemy the king, and sometimes depicted the astronomer wearing a crown (Figure 2–3)— but as all the portraits were acts of imagination, this did not matter very much.

Ptolemy the astronomer collected the 300-year-old observations of Hipparchos, those of astronomers who had worked at Alexandria, as well as some he made himself, and welded these into one scheme of the universe. He was able to do this because he used and developed the deferent and epicycle system of Apollonios. Ptolemy's system, designed to account for every kind of observation, including those that showed that the apparent sizes of the sun and moon varied as they orbited the earth, was extraordinarily complex. For instance, to account for the moon's motion, he had to offset the center of its deferent from the center of the earth and then make the epicycle keep a regular even rotation, not about its center, but about another point near the earth. Even so it was not a perfect solution because it did not account for every observed motion of the moon, but only for the larger and more obvious movements. Ptolemy tried to improve it, but the final solution baffled him, as it has baffled astronomers up to modern times.

For planetary motions, the situation was even more complicated, for not only was the center of the deferent away from the center of the earth, and the rotation of the epicycle regular about a point close to the earth (but differently situated from the one for the moon), but in addition the deferent was tilted one way in space, the epicycle in the opposite way, and then both were believed to rock back and forth. And as if this were not enough, the planet Mercury presented such difficulties that Ptolemy had to consider the center of its deferent performing a circular motion in space too. When one realizes that the future positions of the planets had to be computed by geometry, it is possible to get a glimpse of the immense size of the task.

Yet difficult though Ptolemy's system might be, it was far better than anything before it and did duty for more than 1,000 years. The original Greek text, *The Mathematical Collection*, was known also as *The Great Collection*, but it came to the scholars of Western Europe by way of the Muslim world, as did all the Greek scientific and philosophical writings. On the way, the Muslims abbreviated its title to *The Greatest*, the Greek for which is *megistē*; in Arabic the word *the* is *al* so that in the end, Ptolemy's great work became known as *Almagest*, the title by which we know it today. It was an astounding work, for it contained not only his theory of the motions of the sun and moon, and the planets, his arguments for adopting a fixed earth rather than a moving one (see Figure 2–4), a catalog of star positions, a table

FIGURE 2–4. The earth-centered, or geocentric, theory of the universe. The earth, with Aristotle's four elements—earth, air, fire, and water—is shown at the center of the spheres carrying the sun, moon, and planets, and the fixed stars. The signs of the zodiac are given, and heaven is situated outside the outermost sphere. From P. Apian, *Cosmographia* (Antwerp, 1543). Courtesy of the Ronan Picture Library.

of future planetary positions, but even a section on such basic geometry as was necessary for understanding the theories he proposed.

With the death of Ptolemy sometime after 150 A.D., the development of planetary theory really went into cold storage. The

Muslim world studied astronomy, basing their ideas on Greek texts, and believed for a time that they had discovered a new form of planetary motion, which they called trepidation. This was a rocking motion, but later it had to be rejected because they found it did not really happen. Nevertheless, though they contributed little new, they did keep the torch of learning bright while Western Europe plunged through the Dark Ages, and they handed on Greek astronomy with many useful and interesting comments of their own.

Western Europe began to receive Greek ideas during the twelfth century A.D. Colored at first by Muslim interpretations, many of the original Greek texts gradually became available, and scholars began to argue and to discuss details. Aristotle's adoption of Eudoxos' homocentric sphere theory was known, but the spheres themselves were thought of as real and solid, made from pure transparennt crystal. In fact, the whole Eudoxian theory was revived and elaborated to such a degree that one Italian philosopher, Girolamo Frascatoro, had to use sixty-three spheres to account for new observations made over the previous 1,400 years.

This was an unhappy situation—and a terrifying one for any astronomer who had to compute the future positions of the planets. The old ideas were becoming topheavy and there was clearly a need to stand back from the problem and consider the whole question of planetary motion in a new light.

3 ✳ Gravity

To survey the problem of planetary motion afresh was not possible until the Greek theories had been proven wrong and regarded only as a steppingstone to new ideas. When this did occur, it brought about important changes in man's concept of the universe. These changes still affect us today, so we must spend some time looking not only at what happened, but also at how it happened. For this we shall have to step outside astronomy into physics. But to begin with, let us return to the time when Greek ideas were still an exciting innovation in Western Europe. The crystal spheres and the theory of epicycles and deferents as developed by Ptolemy were described with enthusiasm and discussed with vigor.

By the fifteenth century, however, a perceptible change occurred. Nicholas of Cusa, a pious and learned man who was to become a cardinal, believed that the earth moved. His ideas stimulated others to think about this possibility. The man finally to bring this view to public notice was Copernicus. Copernicus, who was born in 1473 in Torun, Poland, had many interests. He was a cleric in the Roman Catholic Church, a great civil administrator, and a competent physician. For Copernicus astronomy was only an avocation, but he enjoyed the subject. He studied with a teacher who, like Nicholas of Cusa, supported the idea that the earth moved, and Copernicus decided to examine the matter very carefully before coming to a definite conclusion himself. Mathematically, from the point of view of computing the future places, having the sun as center would, Copernicus thought, simplify matters immensely; but before he committed himself, he had to be certain.

After reading as many Greek authorities as he could find, noting that not all had agreed with Aristotle, and finding that a heliocentric theory had been held for a time, Copernicus settled down to observing and computing. His observations were poor in quality, and he was prone to accept previous observations at their face value; nevertheless, he found that he could improve his calculations by using the sun as his center. What is more, Copernicus did not have to "cheat" as he felt Ptolemy had done. Ptolemy's calculations, placing the center of regular motion away from the center of the earth, did not use regular circular motion properly— or so Copernicus thought—but with the sun as center he found this "immoral" feature was unnecessary.

Copernicus was slow to publish, but in 1530 he circulated a manuscript containing the gist of his ideas, and he received considerable encouragement. All the same he might have been content to leave formal publication alone and merely refine his calculations, had he not been visited in 1539 by a young Protestant, Georg Joachim (better known as Rheticus), who came from Feldkirk, Austria (an area known to the Romans as Rhaetia). He stayed with Copernicus for two or three years but after only one year persuaded him to allow a summary to be published. When he left Copernicus sent the entire manuscript to a friend who, after reading it, sent it on to Rheticus. The plan was to have it published under Rheticus' direction in Nuremberg, from where the summary had been issued, but Rheticus moved to Wittenberg as professor of mathematics and the publication was placed under the supervision of Andreas Osiander. Osiander was a Protestant cleric and set great store by a literal interpretation of the Bible, which stated that the earth stood still and did not move in space. In consequence, Osiander added a preface to the book, stating that the idea of the moving earth was really a superior mathematical method for calculating the future positions of the planets, but was not to be taken as proven in fact. He also may have altered the title, but whatever the truth of this, Copernicus could not object either to the title or to Osiander's preface, for when the first copy reached him he was dying.

De revolutionibus orbium coelestium (*On the Rotation of the Celestial Spheres*) appeared in 1543 (see Figure 3–1). It was dedicated to Pope Paul III and caused a considerable stir. It raised a storm of protest among Protestants—doubtless this is what Osiander had anticipated when he wrote his preface—but little from Copernicus' own church. It was an important work and began to make men reconsider their ideas about the heavens, but the acceptance of a moving earth took time. The concept of a moving earth made most headway in Elizabethan England, where it was strongly supported by mathematicians and appeared in popular almanacs and other books for general reading. Owing to the efforts of Thomas Digges who, like his father published an almanac, Copernicus' theory of a sun-centered universe was

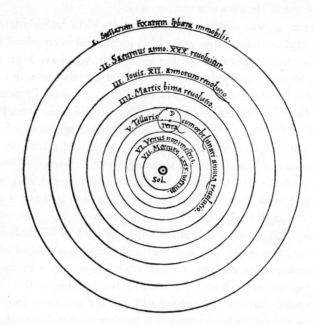

FIGURE 3–1. Copernicus' sun-centered, or heliocentric, theory. From N. Copernicus, *De revolutionibus orbium coelestium* (Nuremberg, 1543). Courtesy of the Ronan Picture Library and the Royal Astronomical Society.

linked with the idea of a universe that extends infinitely into space. This was another revolutionary view, which we shall deal with in more detail in Chapter 9, but it must be mentioned here because of the serious religious conflect that was soon to arise over the new concepts.

The Protestants—or at least some of them—objected to the idea of a moving earth, but in England there was greater freedom of thought, and it was here that Giordano Bruno paid a long visit in 1583. Bruno had been brought up in the Roman Catholic Church and had become a Dominican friar, but he was of a rebellious nature and in 1600 was burned for heresy in Rome. His significance for astronomy is not his fresh outlook, as is sometimes thought, but his acceptance of the Copernican theory as he learned it in England and his incorporation of it in revolutionary books. These, in which he abused his church and tried to advocate a change in its form of government were, naturally enough, condemned, but the tragedy was that they brought condemnation also to the Copernican theory. Thus it was that the Roman Catholic church turned against the heliocentric universe and made it difficult for a man such as Galileo, in Italy, to develop the new theory.

Galileo Galilei, born in Pisa in 1564, was primarily a mathematician and what we would today call a physicist. His contributions to planetary motion and gravitation theory were in physics and mathematics. He brought observational evidence to bear on the question of the earth's movement and discussed the new physical problems that the Copernican theory brought into the open. We shall be considering his observations in the next chapter, and shall therefore confine ourselves here to his theoretical work insofar as it affected the discovery of how the planets really do move in space. Galileo had difficulties in view both of Bruno's behavior and his own fiery character. His book *Dialogo sopra i due massimi sistemi del mondo, Tolemaico e Copernico* (*Dialogue of the Two Great World Systems, Ptolemaic and Copernican*), published in 1632, was later banned and Galileo himself forced to disown the views expressed in it, whereas his later book *Discorsi e dimonstrazioni matematiche intorno à due nuove scienzi* . . . (*Discourses and Mathematical Demonstrations on Two New Sciences* . . .) was published in Holland from a manuscript that, Galileo claimed, had been stolen from him.

Galileo's first book was on his observations and, from the point of view of planetary motion and gravitation the most important of these observations was his discovery that there were small stars—satellites—that continually orbited round the planet Jupiter. One of the arguments against taking the Copernican theory as really meaning that the earth moved was that, if it did, the moon ought to be left behind. What, after all, was there to pull it along? According to Aristotle, it had its natural place in the sky. Of course, one might argue that the earth pulled bodies down so that they fell to the ground after they were thrown up in the air, but celestial bodies were different. They were composed of Aristotle's fifth element. But when Galileo observed Jupiter's satellites in orbit round the planet, the argument against Copernicus lost some of its

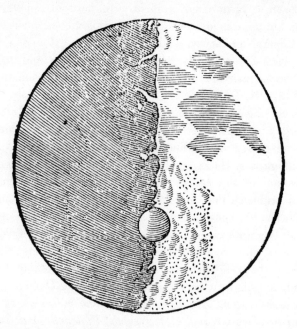

FIGURE 3–2. A drawing of the moon at first quarter, made by Galileo in 1609, showing a crater and mountains. From G. Galileo, *Siderius nuncius* (Venice, 1610). Courtesy of the Ronan Picture Library and E. P. Goldschmidt & Co., Ltd.

force. What is more, Galileo's observations also showed him that, far from being made of celestial material, the moon gave every evidence of having mountains and valleys and being, indeed, just like the earth. Here was evidence that the same laws should apply in the heavens as on the earth. (See Figure 3–2.)

Troubles over the Copernican theory of a moving earth not only revolved around its opposition to some scriptural tests—an argument that in due course seemed to become less pressing—but also around the fact that if the earth moved the whole of Aristotle's physics would be completely upset. Things fell toward the center of the earth because the latter was the center of the universe. But if the earth orbited the sun, why should this be so, for the center of the universe was now occupied by the sun? And, again, if the earth moved through space and rotated on its axis once a day, what would happen to the air? There was nothing to push it, so why should it come with the earth: at best it could only be dragged along and that would mean that we should experience rushing gales and tornadoes. Yet clearly these were exceptions, not the rule. What Galileo realized was that Aristotle's physics needed to be replaced by a better scheme. But he was not the first to think this.

Some dissatisfaction with Aristotle's physics had been growing ever since the middle of the thirteenth century; in particular, attacks had been made on his theory of motion. The trouble really arose from the problem of explaining why an arrow or cannonball (or any other projectile) kept moving once it had been shot into the air. Aristotle believed that any body required a force to act on it continually to keep it moving—just as a cart must be pulled all the time by an ox or a horse if it is not to stop—and he had therefore to find something to push the arrow along from one moment to the next. The string of an archer's bow would start the arrow off, but after that it would be traveling alone in the air. The air, then, must continuously push it. But this involves a constant pressure by one layer of air after another for a time, and then a sudden drop to the ground—for Aristotle and everyone after him thought that a projectile first moved in a straight line up in the air and then fell straight down. This was the only way to substantiate the belief that a body could only undertake one kind of motion at a time.

Objections multiplied and finally, during the early fourteenth century, Jean Buridan, a learned French friar of Paris University, began to work out a new theory of motion. He suggested that a projectile did not require a continual push of air to keep it moving but that, once started, its impetus kept it going. However, he still held to the belief that a body must receive a continuous push if it were to move; in the case of a projectile this push came from the impetus, and the body fell to the ground when the impetus was used up. Others followed this lead, but Galileo really broke the final bonds that tied men to the basic ideas of Aristotle's theory of moving bodies. First, Galileo proved Aristotle wrong over the question of falling bodies. Aristotle had taught that heavy bodies fall to the ground more quickly than light ones, but Galileo found that whatever their weight, they always fall to the ground at the same speed. (There is probably no truth, however, to the story that he made a public demonstration of Aristotle's mistake by dropping bodies of different weights from the top of the Leaning Tower of Pisa, though such an act to confound his opponents would have been very much in character.)

Second, Galileo showed that a body could perform more than one motion at a time, and pointed out that the path of a projectile was a curve (a parabola) not two separate motions in straight lines, one after the other. Third, Galileo made it clear that Aristotle, Buridan, and everyone else was wrong when they thought that a body required a constant push to keep it moving. In his *Discourses and Mathematical Demonstrations on Two New Sciences*, he stated that once a body received a push to give it a certain velocity, it would keep this velocity as long as nothing interfered with it, though he went on to explain that this only occurred in practice when a body moved on a flat surface level with the ground. He did not, therefore, apply this idea to planetary motion, but, as we shall see the principle was to be extended more broadly by Isaac Newton. Nor did Galileo apply to planetary problems the view he held that the change in a body's motion depends directly on the force acting on it. But the fact that it was not he who applied these new physical principles to the age-old question of the movements of the planets does not detract from their importance. Such principles have to be worked out and published before further advances can be made. What is more, for their true value to be appreciated, and before full use could be made of them, the old belief that planets always performed regular circular motions needed to be exploded. By one of those quirks of fate, the astronomer to achieve this was Johannes Kepler, a man whose mind seems in many ways to have remained fixed in the outlook of medieval times.

Kepler was born in 1571 at Weil in the German province of Würtemberg; he was a contemporary of Galileo, who never met him, though the two men did correspond. Kepler first made his reputation by the publication in 1596 of *Prodromus dissertationum cosmographicarum continens mysterium cosmographicum* (*A First Precursor of Cosmological Dissertations Containing the*

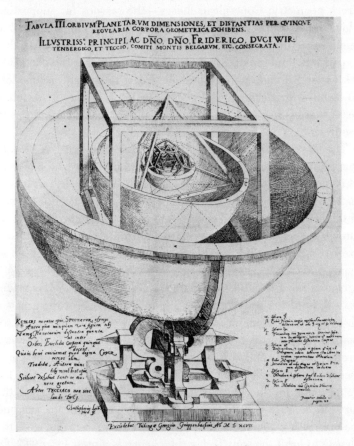

FIGURE 3–3. Kepler's first scheme to explain planetary orbits. It placed the sun at the center, the distance between one celestial sphere and the next being calculated by computing the space needed to fit in one of Euclid's geometrical solid figures with equal sides. This diagram first appeared in *Mysterium cosmographicum*, though this illustration is from the better printed *Harmonices mundi* (Linz, 1619). Courtesy of the Ronan Picture Library and the Royal Astronomical Society.

Mystery of the Universe). The volume contains what today seems a very odd idea, namely, that the distances of the planets from the sun could be determined by seeing which regular shapes from solid geometry fitted in between the crystal spheres, the calculations being based on the time it takes each planet to orbit the sun. (See Figure 3–3.) Kepler's scheme was a curious mixture of Greek geometry, Copernican theory—the reasons for supporting which he expounds very ably—and a kind of number worship reminiscent of Pythagoras and Plato. Yet it is also clear that Kepler was thinking of a force from the sun causing the planets to orbit round it and, more important still, a force that became less powerful with distance. All the same he held the Aristotelian belief that his force had to be applied all the time if motion was to continue.

The *Mysterium cosmographicum* shows how hard it was to change from an old mystical way of thinking to a new one, and why old ideas could not be discarded overnight. It was not a matter of accepting just a new theory, but a completely new way of thinking. Kepler never managed to make a clean break. He was forever seeking a divine inner harmony to the universe. But however strange this quest may seem to us now, with Kepler it led to momentous results.

One of the astronomers who was impressed with the *Mysterium cosmographicum* was the Dane, Tycho Brahe, one of the greatest observational astronomers who has ever lived and whose work we shall be considering in the next chapter. Like Kepler, Tycho was a Protestant. But, unlike Kepler, he did not favor the Copernican theory. The Bible said that the sun stood still, so still it must be. To account for the observational evidence, Brahe assumed the earth to be at rest in the center of the universe with the sun and moon orbiting round it, but he did permit the remainder of the planets to orbit the sun. Today this theory is almost forgotten but during the last years of the sixteenth century and the first half of the seventeenth, it was very popular.

When Tycho read Kepler's book he invited the young man to come and work with him, and because Kepler had to leave his region of Germany owing to religious persecution, he was glad to accept. Tycho was then in Prague, where he died in 1601, and he left his mass of planetary observations to Kepler. It was Tycho's hope that they would prove that his compromise theory was really correct. The observations were the most accurate ever made and Kepler set to work with a will, beginning with studies of the movements of the planet Mars. However, the more he used them for calculation, the clearer it became that Tycho's theory would not fit, and he was forced to discard it. But it also became clear that Copernicus' theory, with its circular motions centered on the sun, would not fit either. In the end, Kepler was forced to break with traditional thinking, and it says much for his determination to be led by observation that he was able to do this. He became convinced that whatever the other planets might do, Mars did not orbit the sun in a circle; it orbited in an ellipse. This kind of oval curve has two points, or foci, instead of one point in the center as we get in a circle, and Kepler found that the sun was at one of these. It took him five years of hard work computing a total of seventy different orbits for Mars (and with no computer to help him) before he reached this conclusion. We can, perhaps, be thankful that his inner compulsion to find a true harmony of the universe was so strong that it kept him at work that others might well have given up long before.

Kepler privately published his results in 1609 in *Astronomia nova "aitiologētos" seu physia coelestis, tradita commentariis de motibus stellae Martis. Ex observationibus G. V. Tychonis Brahe. (A New Astronomy Giving the Cause or Physics of the Heavens, Including a Commentary on the Motion of Mars. From Observations of Tycho Brahe)*—a title that is descriptive enough, if not a model of brevity. But was Mars unique? How did the other planets behave? To answer these questions meant another investigation, calculating orbit after orbit. It took Kepler nine more years, but in 1619 he published the first part of *Epitome astronomiae Copernicanae (Epitome of Copernican Astronomy)*; this was followed in 1620 and 1621 by the second and third parts. The book extended his work from Mars to the other planets, and contained two famous laws of planetary motion: the first law stating now that all planets orbit the sun in ellipses, with the sun at one focus of the ellipse. Kepler's second law was concerned with the velocity with which planets orbit the sun. (See Figure 3–4.)

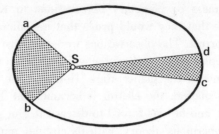

FIGURE 3–4. Kepler's scheme of planetary motion. He discovered that a planet moves in an ellipse (here shown much exaggerated for the sake of clearness), with the sun at one focus S, and at a speed that is faster nearer the sun and slower farther away. A planet will travel from a to b in the same time as it takes to go from c to d, Kepler having found that the line S to the planet covers equal areas (shown dotted) aSb and cSd in the same time.

This, he realized, was not constant, but varied in such a way that a line drawn from the sun to the planet swept out equal areas of space in equal times. As shown in the figure, this means that a planet takes the same time to go from a to b, as from c to d because the area a b S is the same as the area c d S. Why it should do this Kepler did not know, though he did suggest that the sun was like a magnet and exerted a magnetic force on the planets.

But in spite of these achievements, Kepler was dissatisfied with his basic work. He wanted to find underlying the universe, some mathematical relationship that would link all the planets together in a vast embracing scheme of beauty. At last in 1619, while his *Epitome* was being published, he produced another book extending the work he had originally done on planetary distances and given in his *Mysterium cosmographicum.* This new volume, *Harmonices mundi (Harmony of the World)* was, he thought, his crowning achievement, because it contained details of an elaborate relationship he had worked out between the velocities of the planets when nearest to and farthest away from the sun in their elliptical orbits—their speeds at perihelion and aphelion. Kepler had found that the relationship between the velocities would give fractions and that these fractions were similar to those obtained when one compared the intervals between musical notes. From this he worked out a kind of music of the heavens—a divine harmony. All this had taken much computing, during which

Kepler found a third relationship between the planets, a relationship that is known now as his third law. It states that for any two planets, the square of the time taken by one to orbit the sun, divided by the square of the other's orbital period, is equal to the cube of the average distance of the first planet divided by the cube of the average distance of the second planet.[1] In other words it was a law connecting planetary distances with the time they take to orbit the sun, and its use did not depend on knowing how many miles away a planet lay from the sun, but only on a comparison of planetary distances. For example, if we take the period of the earth's orbit to be one (one year) and its average distance from the sun as one (one astronomical unit, as it is usually called), then if we observe that another planet, Jupiter, say, takes 11.8 years, the law allows us to work out that Jupiter is at an average distance of 5.2 astronomical units, that is, slightly more than five times the distance sun to earth. Unfortunately this third law was wrapped up in, and secondary to, Kepler's doctrine of divine harmony, and this was to present a problem in later research.

Kepler thought of the sun exerting a magnetic influence on the planets, less strong the farther away a planet lay. But he was not alone in considering a force of this kind, though few followed him in believing it to be magnetic. Giovanni Borelli, who was only twenty-two when Kepler died in 1630, rejected both Kepler's magnetism and also his suggestion that the sun sent out a force that acted like the spokes of a wheel, pushing the planets along their orbits. After much work, which included a careful study of the behavior of Jupiter's satellites, he concluded that three forces operated to keep the planets in their correct paths. First, there was the planet's appetite for the sun, second, an opposite force tending to throw the planet off in space directly away from the sun, and, third, a force lying along the planet's orbit that caused its motion. Still, like Kepler and others before him, Borelli thought a planet

[1] Expressed in mathematical terms

$$\frac{P_1^2}{P_2^2} = \frac{a_1^3}{a_2^3}$$

where P_1 and P_2 are the periods of two planets, and a_1 and a_2 their average distances from the sun.

required an unending push to keep it moving, but his idea of a planet's appetite was widely discussed. If there were such a force, what kind was it? If it altered with distance, how did it alter?

Matters were explored further in England, where a number of scientific men, members of a new scientific society, the Royal Society, used to meet once a week to discuss scientific problems. Among those who talked over the question of a planetary appetite or gravitation (the word was often used) operating from the sun, were the architect and astronomer Christopher Wren, a young astronomer Edmond Halley, who had recently mapped the stars visible in the southern skies, and the Royal Society's "curator" of experiments, Robert Hooke. They came to the conclusion that there must be a force of gravitation operating from the sun, but how did it vary with distance? Perhaps it followed an inverse square law, that is, became only one quarter as strong at double the distance, one ninth as strong at three times the distance, and so on.[2] This would certainly be reasonable—indeed it was quite likely—but it was difficult to prove. Wren could not do so, and the mathematics defeated Hooke and even Halley, who was known for his mathematical ability. Yet both Hooke and Halley tried hard, for Wren had offered an expensive book as a prize to the first to solve it.

Achieving no success himself, Halley decided in August 1684 to go to Cambridge to discuss the matter with a professor of mathematics, Isaac Newton. Then aged forty-two, Newton had gained a great reputation and was not only able to confirm that the gravitation force did follow an inverse square law, but also claimed that he had proved the point mathematically. He could not find the proof he had once written out, but promised to send a copy on to Halley in London. Halley returned and eagerly awaited Newton's proof, which arrived three months later. Halley was mathematician enough to appreciate the immense significance of Newton's thinking about motion, and the outcome of his reports and persuasions was that Newton agreed to write his results down for the Royal Society to publish.

[2] In other words, a force that varied inversely as the square of the distance, or $1/d^2$.

Newton's work on gravitation and planetary motion had begun in 1665 when he had to leave Cambridge University because the Great Plague had reached the city, probably from London. He had returned to his home in the village of Woolsthorpe near Grantham and here, in the peace and quiet of the countryside, had tackled the problem. It is said by his niece that, while sitting one day under an apple tree in the garden considering the moon's motion, an apple had fallen with a thud at his feet, and led him to wonder whether the force that pulled the apple to the ground was the same as that which pulled the moon toward the earth as it performed its monthly orbit. He had calculated the matter out, assuming an inverse square law, and found that his answer was promising. But what of the planets?

Newton pursued his calculations, using a new mathematical method that he had invented, which we call calculus. He had difficulties, however, for he had to determine whether the pull of gravity of the earth operated as if it were a force acting from the center of the earth and not from the surface. He had to work with inexact values for the distances of the moon and the size of the earth, and it now seems likely that he did not know about Kepler's second law. This law concerns the way a planet travels across equal areas in equal times when its motion is measured from the sun. It appears that a number of scientists did not know of it, and even those who did were discouraged by his discussion of its musical relationships and the lack of mathematical proof.

By 1684 Newton had solved his problems and, with encouragement from Halley, set about writing down the whole subject. In doing so, Newton decided against using his new calculus because it would be too little understood; if readers had to grapple with that before tackling the difficulties of planetary motion, they would never read what he was to write. Halley was editor and paid the cost of printing, but only after many troubles—including an argument between Hooke and Newton that nearly left the book unfinished—the volume was published in July 1687. *Philosophiae naturalis principia mathematica* (*The Mathematical Principles of Natural Philosophy*) has been said to be the most important scientific treatise ever written. This is an exaggeration, but even so the *Principia* marked a turning point in man's study of the universe and, in fact, of all physical science.

In the book Newton laid down three laws of motion, the first and second being extensions of Galileo's work to which, one is glad to see, he gives due credit. The first, which the French philosopher René Descartes had also found, states that a body will continue either in motion or in a state of rest forever, unless some force acts on it. This was the final break with Aristotelian tradition, and it meant that it was no longer necessary to imagine that the planets needed a constant push to keep them going.

Newton's second law concerned change of motion, which, he explained, is owing to the strength of the force acting on a body, and occurs in the direction in which the force operates. The third law states that when two bodies act on each other, their reactions are both equal but in direct opposition. These last two laws are also important when it comes to working out planetary behavior using gravitation. Gravitation was at last given a precise mathematical meaning in the *Principia*. It was a force acting between bodies, its strength depending on the massiveness of the bodies concerned and their distance apart. It weakened with increasing distance according to an inverse square rule just as Halley, Wren, and Hook had supposed. With these basic laws, Newton then proceeded to work out the motions bodies would take under various forces, provided they were moving in empty space. He continued by discussing such movements when bodies moved through something that resisted their motion and, finally, in the last part of the book, he applied all these results to the movements of the moon and the planets. All the observed motions were explained with a degree of precision never attained by any other theory. (See Figure 3–5.)

The scientific world was amazed by Newton's achievement, and his theory was accepted, except by some scientists in European countries who were unhappy about the force of gravity. Where did it lie—in a body or around a body? What was it? How did it travel through space, or was it acting at a distance? Newton could not answer these questions—all he would say was that if one supposed that there was a force of gravity and that its strength was as he had described, then when it was applied to planetary motions or to the motions of bodies on earth, it explained the facts. But for those who felt that he was really proposing a mystical force such as a medieval astronomer might have done, this argument did not seem sound. What was really needed was some independent proof of Newton's theory.

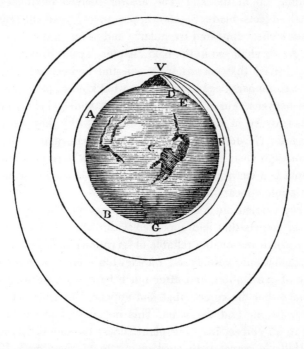

FIGURE 3–5. One of the consequences of Newton's theory was the behavior of a projectile shot off from the earth. This drawing, which appeared in Newton's *The System of the World* (published in London in English in 1728), shows various paths that a projectile could take. Each path depends on the velocity with which the projectile launched; when the velocity is great enough, it will go into orbit around the earth (path VAB) as an artificial satellite does. Courtesy of the Ronan Picture Library.

It is pleasant to record that the man to achieve this first was Edmond Halley, Newton's friend, without whose efforts the *Principia* would never have appeared. As soon as the *Principia* was published, Halley set to work on the question of the paths of comets. In earlier times, comets with their bright heads and long fiery tails were thought of as bringing evil, and looked on with terror. Aristotle had taught that they were in the upper air, and the idea that these apparently hot fiery things could bring dry conditions suitable for plague and pestilence was not so very farfetched. Why could they not be astronomical objects instead of something up in the air? The answer seemed obvious—astronomical objects made regular appearances, year in year out, whereas comets appeared irregularly and never in the same place twice. As we shall see in the next chapter, Tycho Brahe, in 1577, finally proved that comets were celestial bodies, and after this many astronomers began to consider what kind of paths they took in the heavens. For a time there was a widely held view that comets traveled in straight lines, which was why they were never seen again. Newton puzzled over the problem, and in the *Principia* made it clear that he believed the paths were parabolas; the parabola is a curve open at one end, and this would mean that a comet would appear once only.

Halley carried out a thorough investigation. He computed the paths of comets that had appeared within the last few centuries, during which reasonably reliable observations had been obtained. His calculations were based on Newton's laws of motion and theory of gravitation, and after much hard work he was able to announce that the comets that had appeared in 1531, 1607, and 1682 were one and the same. This meant that the comet must move in an ellipse, just as a planet does; the one difference was that Halley's comet path worked out to be very oval. But not content with this, he calculated when the comet should next appear, and found it to be late in 1758. Halley published his views in 1705, but died in 1742, not knowing whether his prediction was correct. The comet appeared on Christmas Day 1758 and both Halley's prediction and Newton's theory were magnificently vindicated.

The accuracy of Halley's prediction removed any lingering doubts, and many more mathematically minded astronomers began to adopt Newton's theory and investigate lunar and planetary motions in more detail. Because of an unfortunate argument that Newton had with the German mathematician Gottfried Leibniz about who invented the calcus—both Newton and

Leibniz had done so independently—British astronomers refused
to use Leibniz's methods and adopted Newton's, which were more
complicated. Consequently, after Halley had completed his
cometary work, little progress in developing Newton's theory was
made in Britain for some time, and it was left to mathematicians
in France and Germany to work out the gravitational influences of
one planet on another. Many were involved in this intricate work,
most notably Pierre Laplace and Leonhard Euler.

The applications of Newton's theory to planetary problems did
not end with the work of Euler and Laplace. During the 1840's
some doubts were cast on it because the planet Uranus, discovered
by William Herschel in 1781, was not following the orbit it should
under the influence of gravitation. There seemed only one way to
save the situation, and this was to assume that there must be yet
another planet, farther out than Uranus, affecting Herschel's
planet by its own gravitational pull. To compute where such a
planet should be and how massive it must be to account for the
disturbances suffered by Uranus, was a difficult task—it has been
said to be equivalent to playing chess on a three-dimensional
board (where the pieces can move up and down as well as forward, backward, and sideways) with a blindfold over one's eyes!
However, two mathematicians tackled the problem—Urbain
Leverrier in France and John Adams in England. Both were successful and both obtained the same result, and when in 1846
observations were made of the place in the sky they had predicted,
another planet was seen—Neptune. Yet after some years this, too,
showed irregularities. But by then Newton's basic theory was too
well established to be cast aside—even though its principles had
been much extended in 1915 by Albert Einstein's theory of relativity—and yet another planet, was computed to be the cause. The
computations followed the lines of those of Adams and Leverrier
and were mostly carried out by the American astronomer Percival
Lowell, who searched for the planet from the observatory he had
established in Flagstaff, Arizona. But Lowell had no success with
his observations and when he died in 1916 the quest for the planet
was dropped for a time. William Pickering, another American,
took the matter up again during the 1920's, and when he too
achieved mathematical results similar to Lowell's, astronomers at

the Lowell Observatory again set about searching for the supposed planet. A year later, in 1930, Clyde Tombaugh found it close to the calculated position, using photographs to aid him in his discovery.

Pluto's discovery seemed another proof of Newtonian principles. But there was still an additional case of difficulty, and this was the behavior of Mercury. The rather oval orbit of this planet was found to rotate but at a speed greater than that which Newtonian gravitation would lead one to expect. This fact was appreciated by Leverrier, and he computed an orbit for a disturbing planet nearer to the sun than Mercury. He announced the results of his calculations to the French Académie des Sciences in September 1859, and showed that the effects on Mercury's orbit could be caused by a planet of Mercury's size orbiting the sun at half its distance. Following directly on this announcement, an amateur astronomer, a country doctor at Orgères, some fifty-five miles southwest of Paris, claimed that he had already observed the new planet six months earlier. Leverrier hurried to see him and claimed that the observations were the proof he needed; he named the planet Vulcan. Though subsequent claims of successful observations were made by professional astronomers, a careful check in March 1877 and October 1882, when Vulcan should have been observed moving across the face of the sun, showed nothing. Astronomers do not now accept Leverrier's conclusion, and they no longer believe in Vulcan's existence. Instead, they think the true explanation of the behavior of Mercury's orbit is to be found by applying Einstein's theory of relativity, which accounts for the movement almost exactly.

With the coming of Newton's theory of gravitation, the main problem of planetary motion was solved. Though Laplace, Euler, and others worked out its consequences in detail, and Einstein's relativity theory extended it, the fact remains that Newton's work closed one long section in the history of astronomy. But it also opened a new one, inasmuch as Newton had applied precisely the same laws of physics to things happening on earth and in space. This was the final break with Greek ideas, which had dominated astronomy for so long, and ushered in a new outlook, to become evident in future chapters.

4 ✻ Measurement of Position

As we saw in Chapter 1, one of the great difficulties in forming a precise picture of the universe is to get an idea of scale. How far away is the moon? How far off is the sun? Which of the two planets Mercury and Venus is nearer? Are the stars on a sphere, or do they lie at different distances? These and many other questions face the astronomer, and they must be answered if we are to form a picture that is anywhere near correct. But the answers involve us in measurement—very careful measurement—which is such a basic need that astronomy is classed as one of the exact sciences, and some consider it the most precise of all.

The earliest measurements were rough by modern standards, but by the standards of the time they were highly accurate. They fell into two main categories: measurements of time, to provide details of when something happened, and measurements of angles, to specify where it occurred. The history of the measurement of time we shall trace in Chapter 5, because it will be more convenient if we begin with the measuring of angles. To use angles may, on the face of it, seem a strange way to measure position, but considering that the stars and planets all appear to be the same distance away (whether or not they are does not concern us at the moment), it is quite reasonable to think of them spread out on the inside of a sphere. And if they are spread out on the inside of a sphere, angles are the most convenient way to specify position.

Angle measurement was adopted by the Babylonians, whose method of dividing a circle into 6 × 60, or 360° we still use. For fractions of a degree, the sexagesimal system was also followed, so that we have sixty minutes of arc in a degree, and sixty seconds of arc in a minute. But when we reach fractions of a second—as happens in modern measurements—astronomers adopt the decimal system of tenths, hundredths, and thousandths. Also, they now find minutes of arc or seconds of arc too long to say, and refer to them as minutes and seconds, writing them ° for degrees, ′ for minutes, and ″ for seconds; this is a convenient form of shorthand.

The earliest observers must have used very simple apparatus, but few details have reached us. We do know, however, that the Egyptians used a merkhet (see Figure 4–1), which consisted of a

FIGURE 4–1. The merkhet, using a split stick, plumb line, and a tree to obtain the moment of meridian passage of a star (that is, to determine when it is due south).

stick with a slot dividing part of it, and a plumb line (a string with a weight on the end to make it hang vertically); sometimes there were two plumb lines. The purpose of the merkhet was to observe the moment when a star passed due south, and so the split stick and plumb line were kept aligned with a tree or some similar object (or the second plumb line), so that the observer peering through the split stick was looking due south. He would then be able to call out to one of his assistants the moment the star was in line with his plumb line.

Two other very simple instruments were the gnomon and the plinth. The earliest gnomon was a vertical stick placed in the ground; due south could be obtained by noting the direction of the shortest shadow cast by the sun, since the shadow is shortest when the sun is at its highest, and this occurs when it is due south.

The plinth was essentially no more than a narrow block of wood or, more frequently, stone laid with its narrowest sides facing due north and south. Carefully leveled by putting small wedges underneath, it could indicate the height of the sun when due south, either by the shadow cast from a peg protruding from one of the sides, or else from the length of the shadow cast by the block itself on a central opening. Measures of altitude could also be made using a vertical ring fixed so the observer looked in a southerly direction (Figure 4–2 a)—the meridional armillary—and he could also obtain angles along the celestial equator from a specially tilted ring (Figure 4–2 b.)

FIGURE 4–2. Meridional armillary (vertical ring—a) and equatorial armillary (b), two fixed types of observing instrument without moving parts.

As observing continued, astronomers wanted to plot the positions of the stars and planets with increasing accuracy, and this meant more elaborate instruments, that is, instruments with moving parts. A basic quite elaborate device of this kind was known as Ptolemy's rules, though it was almost certainly not invented by Ptolemy and was probably used at least as early as the time of Hipparchos. It consisted of a post that could pivot about a fixed post and was set vertical by a plumb line. Two wooden rods were attached to the movable post, one at the top and one at the bottom. Each was pivoted and they connected with each other.

One arm contained open sights along which the observer could peer so that it could be lined up with a star or planet. The arm with the sights was laid on the object being observed, and where it was crossed by the other arm a figure could be read off; this gave the elevation of the celestial body.

Once one instrument with moving parts had been constructed, others followed; these took various forms. Essentially they were all devices with open sights and arms or disks engraved with numbers that gave the observer the position of a celestial body. Some of these were beautifully made, the craftsmen who constructed them trying to build instruments that were elegant and would delight their rich patrons. (This is one of the reasons why they are so highly prized by collectors today and take their place among the minor art treasures of the world.) Three types are worth noting because they were so popular and remained in regular use until the close of the seventeenth century. The oldest is probably the instrument that Ptolemy called an astrolabon, known now as an armillary sphere. It consisted (see Figure 2–3) of a number of rings, each mounted on a pair of pivots, and one of the rings contained the open sights. To use it the observer laid his sights on the celestial body by rotating the various rings. He then read off the position of the body from numbers engraved on the rings. Ptolemy used an instrument like this for his planetary observations and, especially, for his observations of the moon.

A variation of the armillary sphere was known as the torquetum, which made use of disks instead of rings. Both were very convenient for determining the positions of the moon and the planets because astronomers measured these in degrees north or south of the sun's path in the sky—the ecliptic. Distances along the ecliptic were measured in degrees, beginning at the point where the ecliptic crosses the celestial equator (Figure 4–3).

The other widely used astronomical instruments were the astrolabe (which had nothing to do with Ptolemy's astrolabon), the cross-staff, and the quadrant. The quadrant consisted of a sighting arm mounted on a pivot, and its position was read off against a scale engraved on a metal plate or painted on wood. As its name explains, the quadrant could only be used for measuring angles up to $90°$, or a quarter of a circle ($360°$). The astrolabe was more elaborate, having a longer scale and being a simple computer as well. It was probably invented some time after the sixth century and was shaped like a flat metal plate. One side of the metal plate was engraved with degrees (and fractions of degrees) and the other with various lines. The sighting bar lay over the side containing the degree scale, and a rete, or net, of metalwork lay on top the other side. When the disk was turned, the arms of the rete acted as pointers to provide answers to questions such as where the sun was relative to the stars; the disk and rete together formed

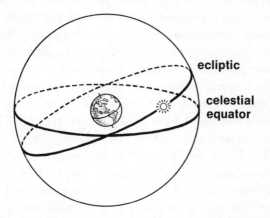

FIGURE 4–3. The celestial equator, which is in line with the earth's equator, and the ecliptic along which the sun appears to move.

a moving and self-calculating star chart. To observe with the astrolabe, one held it by a ring at the top, and it could be used anywhere—on land or at sea. The cross-staff was a wooden device used for measuring the angle between two celestial bodies. The observer placed his eye at one end of the staff, and moved a wooden arm lying across the staff until it seemed to fit between the celestial bodies. Their angular distance could then be read off from a scale cut in the long arm. Like the astrolabe, it could be used anywhere, and navigators, as well as astronomers, gradually found it indispensable.

None of these instruments was particularly accurate, and measurements could only be made correct to 20′ unless very special precautions were taken. An accuracy of 20′ is not very precise; it means that one could only fix a star somewhere within an area of the sky a little smaller than the diameter of the moon and was inadequate for determining planetary motions in great detail. It was also far too large an error for any distance measurements of stars to be successful. All the same, the early observers could do no better, and a degree of accuracy close to this had to satisfy Hipparchos, Ptolemy, and others of late Greek times who prepared catalogs in which star positions were listed. One way of trying to obtain increased accuracy was to construct instruments of immense size with huge scales marked on them. Thus it was that very large instruments, built of stone and brick, were made by Muslim astronomers (see Figure 4–4) and, later, by the Chinese. The first appears to have been a quadrant, with the scale on a curved masonry wall and a peephole in a vertical wall at one end. The peephole could act as an aperture for sunlight or moonlight, and was about twenty-two feet above the ground. The instrument was built by the Muslim astronomer Abūl Wafā about 995 A.D. Larger instruments of this kind were constructed in later years, and Ulugh Beg, a grandson of the famous warrior Tamerlane, whose observatory was in Samarkand (now in Uzbekistan in the south of the U.S.S.R.), had a quadrant whose vertical wall was 180 feet high, and in his star catalog positions were given correct to 10′. Nowadays, the best known of these observatories is at Jaipur in India, but this was built about 1728 by Jai Singh, long after such observing instruments had become out of date.

The trouble with very large instruments was that they were cumbersome, and an accuracy of 10′ was all that could be achieved. By the sixteenth century at least, astronomers wanted to do better than this, and the Danish astronomer Tycho Brahe managed to reduce errors to as low as 4′. Tycho obtained this accuracy with smaller instruments, even though he did have one brickwork quadrant. His success was owing partly to brilliant design, and partly to his ability as an observer. King Frederick of Denmark gave him the island of Hven, in the sound about fourteen miles north of Copenhagen and opposite Elsinore, the scene of Shakespeare's *Hamlet*. Here Tycho established two large observatories near to each other and employed his own workmen to construct his instruments. One of these is shown in Figure 4–5;

FIGURE 4-4. Observatory at Benares, India, using large astronomical instruments built of stone. From an eighteenth-century copperplate engraving. Courtesy of the Ronan Picture Library.

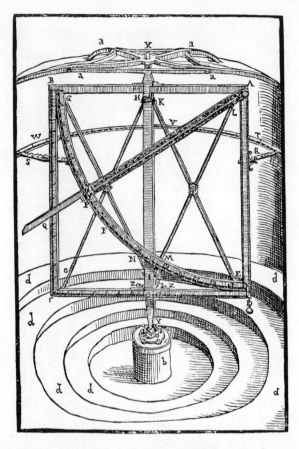

FIGURE 4–5. A square framed metal quadrant, mounted on a pivot and sunk a little way into the ground, with a surrounding shield to prevent the wind from rocking the apparatus and so causing errors in measurement. Designed and used by Tycho Brahe at his observatory on Hveen, and published in his *Astronomiae instaurate mechanica* (Prague, 1602). Courtesy of the Ronan Picture Library.

they were made with every care, and their scales were engraved as precisely as possible. Yet, even so, Tycho appreciated a fact that experimental scientists have taken to heart ever since—however carefully constructed an instrument may be, it can never be free from every error; there will always be some slight imperfections. Tycho therefore made his observations in such a way that he could reduce the errors of his instruments to a minimum.

As we saw in the previous chapter, Tycho's observations led to such accuracy in determining planetary motions that Kepler was able to derive his three laws of planetary motion from them. But before this important discovery was made, Tycho himself had put his art of observing to good advantage.

In 1572 a bright star suddenly appeared in the constellation Cassiopeia. Known then (as now) as a nova, or new star, general opinion placed the object fairly close to the earth, on the argument that it was a change in the heavens and, as such, must lie nearer the earth than the sphere of the moon; the heavens beyond the moon's sphere were, as Aristotle had taught, unchangeable. Tycho decided to examine all the observations of any value—his own as well as those of other such astronomers as Thomas Digges, on whose care and accuracy he could rely. He then compared the observations made from different places in Europe, and in this way was able to determine the distance, or parallax, of the nova.

The method Tycho used was already well known in surveying on the earth (see Figure 4–6). The surveyor first observes from A,

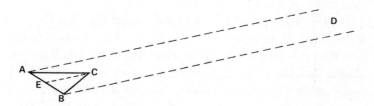

FIGURE 4–6. The principle of obtaining stellar parallax. During a year the earth moves from one side of its orbit at A to the other side, B. A star at C appears to shift its position with reference to a more distant star D. $\angle$ DAC and $\angle$ DBC allow $\angle$ ECA to be calculated, and the distance EC to be computed.

then from B, carefully measuring the distance between the two points. His observations consist in noting the angle between C (the object whose distance he is trying to find) and some far more distant point D. This angle is different when observed from A and from B, as you can see for yourself if you hold up a finger at arm's length and peer at it first through one eye (A) and then the other (B). Your finger (C) appears to shift against the more distant background, and if you took the trouble to select some object (D) on the wall, you could measure the different angles between your finger (C) and the object when viewed from each eye. The angles ($\angle$ DAC and $\angle$ DBC) allow the surveyor to calculate the angle at C ($\angle$ ACB) and, knowing the base line distance AB, the angle allows one to calculate the distance CE (shown by a dotted line). $\angle$ ECA is the same as $\angle$ ECB and half $\angle$ ACB; because it is in direct proportion to the distance CE it is called the parallax of C.

By taking observations from observers at known places in Europe, Tycho was able to try to work out the parallax of the nova, but his results did not enable him to obtain any definite distance—they were too small. But what he was able to discover was that the nova's parallax was certainly much less than the moon's, and this meant that it was much farther away and undoubtedly a truly celestial object. In 1577 a large comet appeared. It was widely observed, and Tycho had plenty of observations to use when he tried to find its parallax. Again he could not obtain a definite answer in terms of distance, but he was able, once more, to prove that it was a celestial body and not something in the upper air as Aristotle had believed.

The fact that Tycho could not obtain a real value for the parallax of either the nova of 1573 or the comet of 1577 was owing, we now know, to their being so far away that their parallaxes were smaller than the errors he obtained in his observing (4'), but he could obtain the moon's parallax because it is so much larger (more than 57' on the average). Some improvement on Tycho's accuracy was obtained in the next century by a rich brewer and outstanding observer, Johannes Hevelius, who built an observatory on the roof of his house in Danzig in the north of Poland. Hevelius' instruments were even better than Tycho's, and, in addition to taking account of their inherent errors, he used two observers working together to measure the angles between stars. The idea behind this was that inasmuch as the stars are continually moving across the sky (owing to the earth's rotation), they will shift in position between the time an observer lays one of his sights on one star and the second sight on another. His delay will produce a quite noticeable error. Hevelius designed his equipment so that each of the two sighting bars was laid on its particular star at the same moment, but two observers were needed for this, one on each sighting arm. Hevelius claimed an accuracy twenty-four times greater than Tycho's (10″), but this was probably an exaggeration. Yet even were it true, it would not be good enough to determine the distance of stars though it would have been able to deal with the 1577 comet. But Hevelius' accuracy was unique and could not be approached by any other astronomer using instruments with sighting bars. It could, however, be equaled by measuring with instruments that had telescopes fitted to them, but this was a point over which Hevelius was extraordinarily pig-headed. He used telescopes for observing details on the moon and planets (see Figure 4–7), but steadfastly refused to apply them to his

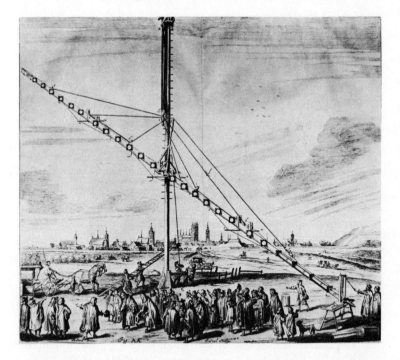

FIGURE 4–7. A refracting telescope more than 200 feet long. Used by Hevelius, the instrument was difficult to handle and almost impossible to use except on a very still night. The immense length was needed because of the poor quality of early lenses. From J. Hevelius, *Selenographia* (Danzig, 1647). Courtesy of the Ronan Picture Library and Royal Astronomical Society.

measuring equipment. With so keen an eyesight as Hevelius must have possessed, there is probably some excuse for his attitude, but it is a pity that he would not change his mind, because the results he could then have obtained might have been even more surprising.

No one has yet been able to discover who invented the telescope, and the whole of its early history is obscure. The friar Roger Bacon, who lived in the thirteenth century, had used magnifying glasses, and suggested that lenses could make distant objects appear closer, while in the sixteenth century Leonard Digges (Thomas Digges' father) had written about surveying and mentioned the use of something that sounds very like a telescope. But whatever the final answer may turn out to be, we do know that in 1608 three Dutch spectacle-makers claimed to have built the first commercial telescopes, and credit is usually given to Hans Lippershey, who was one of them. The instruments were not of very good quality from an optical point of view, but they did show distant things closer and stimulated the interest of Galileo. Galileo who, according to his own account, might well have become an artist instead of a mathematician, was an expert in perspective, and this meant that he had studied the behavior of rays of light. Consequently, when news of the manufacture of telescopes reached Venice in the summer of 1609 Galileo, knowing only that the instruments had two lenses, was able to design and build one for himself. It was a very imperfect instrument, but he soon improved it and, in due course, employed an optician to make larger lenses for him. Galileo used the telescope not only to observe the heavens and find what to him was clear proof of the error of the Aristotelian universe but also for measurement. In his case, however, the measurements were not of star positions but of the height of mountains on the moon. His values were rather too large, but all the same he laid down the principle of such measurements, which was to observe the extent of the shadow cast by the mountains and then work out their height from a knowledge of this and of the sun's position above the moon's surface.

Galileo announced his telescopic observations and gave some details of the design of his instrument in a small book, *Sidereus nuncius (The Sidereal Messenger)*, published in Florence in

1610. Galileo's design of the telescope allowed the observer to see only a very small portion of the sky at any one time, but, in spite of its limitations, it awoke immense interest. Kepler improved the design somewhat and obtained a wider view of the sky, publishing his method of construction the next year, 1611, in his *Dioptrice* (*Dioptrics*—the study of lenses and the refraction of light). The difference between Galileo's and Kepler's designs lay in the lens near the eye—the eyepiece—and both are shown in Figure 4–8.

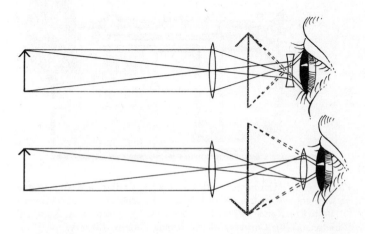

FIGURE 4–8. Galileo's and Kepler's designs of refracting telescopes.

The Keplerian type of telescope was more widely adopted. To begin with the telescope was used for straightforward observing, but gradually it was realized that it offered advantages for measuring the positions of celestial objects. The reason for this is not difficult to appreciate. For instance, imagine that you are looking at a ship coming in to dock. When it is a long way off you cannot see the name painted on the bows because the letters are too close together to be picked out and seen for what they are. But if you use a telescope, you can see them because they seem larger and you can resolve them into separate letters.

Various devices were placed at the eyepiece of a telescope to aid accurate measurement. Christiaan Huygens in Holland used a tapered metal strip to cover a planet's disk exactly and measured

its apparent size from the width of the strip. But it was William
Gascoigne in England who, in 1640, designed and built the first
micrometers (measuring devices) that had more than one moving
part. In one he used two pointers, and in the other two strands of a
spider's web stretched across a metal frame; in both cases the
positions of these could be read off a dial (Figure 4–9). Gas-

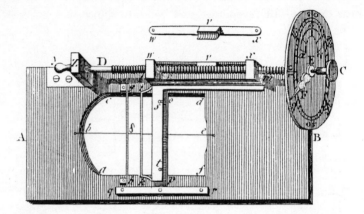
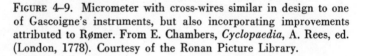

FIGURE 4–9. Micrometer with cross-wires similar in design to one
of Gascoigne's instruments, but also incorporating improvements
attributed to Rømer. From E. Chambers, *Cyclopaedia*, A. Rees, ed.
(London, 1778). Courtesy of the Ronan Picture Library.

coigne was killed during the Civil War in England at the battle of
Marston Moor (1644), but the device was independently designed about the same time by two Italians, Francesco Generini
and Geminiano Montanari. In 1666, Jean Picard and Adrien
Auzout, two French astronomers who were to become famous for
their astronomical measurements, reinvented the device, again
using two strands of spider's web, and in 1672 a Danish astronomer, Olaus Rømer, who came to work with Auzout and Picard,
invented yet another micrometer; his design was widely adopted.
The use of threads from a spider's web may sound almost un-
believable, especially today when very fine metal wires or lines
engraved on glass are usual, but it is perfectly true. Spider's
threads are free from the lumps and fluff that one gets with cotton
or linen thread and are much finer. I have actually seen an instru-

ment mechanic, who was trained in the latter part of the last century, split a spider's thread using a razor blade!

Picard and Auzout developed two other important measuring devices—the zenith sector and the transit instrument. The zenith sector consisted of a small telescope that pointed straight upward (to the zenith or overhead point) and was pivoted near its front end. The observer lay on his back to look through it. There were cross-threads in the eyepiece, and the telescope was moved to line them up with a star. The telescope's position was then read from the scale on the sector and the time of observation noted. This procedure was repeated for another star close to the zenith, and from the readings on the sector and the times the angle between them could be calculated. The transit instrument was a small telescope set up so that it pointed due south (to the meridian) and, using the cross-wires in the eyepiece, the moment when a celestial body crossed the meridian could be timed. This method of timing for determining position led to the celestial sphere being divided into hours, minutes, and seconds of right ascension. The transit instrument was very useful in determining star positions, especially in its later form of meridian circle, where the elevation of a star above the horizon could be measured as well. The transit instrument was developed as the years went by, but the zenith sector was hardly changed for some time.

Now that better instruments had been developed, some astronomers were anxious to see what they could do with the problem that had defeated Tycho—the measurement of stellar distance (parallax). With this in mind James Bradley and Samuel Molyneux improved the zenith sector and settled down to what they knew might well be a long, tedious series of observations. They chose a bright star called γ (gamma) Draconis, which appeared close to the zenith over London (from where they were observing), because it was bright and hence more likely to be near than a dim star. They wanted a zenith star because starlight from overhead has to pass through less of the earth's atmosphere than starlight from closer to the horizon and, as the atmosphere distorts the measured position of a star, they wanted it to have as little effect as possible. Molyneux commissioned George Graham, a well-known instrument-maker of the day, to build him a zenith telescope of great length. The top pivot was fitted to a chimney of the house on his estate at Kew, a few miles southwest of London, and the telescope was passed down through the roof and floorboards to the observer twenty-four-feet below. (Molyneux clearly considered his astronomy more important than his home!)

The instrument was ready in December 1725, and Molyneux began observing with Bradley as assistant. Hooke had observed the same star more than fifty years earlier and thought he had detected a small parallactic shift against the background of the other stars, so Molyneux and Bradley had high hopes of some valuable results. In no more than fourteen days they noticed a shift, but was this shift owing to parallax? If it were, then it was so noticeable that the star must be nearer than expected. The results became more puzzling after further observation, for the greatest shifts appeared in March and September; yet, if the shift were owing to the motion of the earth in its orbit giving rise to parallax, then the greatest shift should be in June and December for the star they were observing. Something seemed to be wrong: Bradley pursued the question. The greatest shifts were six months apart, but the wrong six months, and only in 1728 did he hit on the explanation—the shifts resulted from the velocity of light. His discovery was published the next year in the *Philosophical Transactions of the Royal Society*. The velocity of light had been measured in 1675 by Rømer, who observed the intervals between the eclipses suffered by one of Jupiter's satellites and noticed that these intervals were larger when the earth had moved farther away in its own orbital motion. From these observations he calculated that the speed of light was 120,000 miles per second (186,300 miles per second is the present-day value), but few believed him, and fewer still thought that light took any time at all to travel. Bradley's observations of γ Draconis confirmed that light had a definite velocity, for he showed that what was happening was that as the light passed down the tube of his telescope, the moving earth carried the telescope along so that the light ended up at the side of the eyepiece instead of at the center. This gave an apparent shift of the star's position seen through the telescope, the amount of the shift depending on how the earth was moving relative to the star; that is, where it was in its orbit.

The discovery, known as the aberration of light, was important because it had to be taken into account if a true value was to be obtained for any measure of parallax. It was important also as the first observational proof that the earth really did move in space. For the previous 180 years astronomers had talked about and accepted the idea of a moving earth, but they had no proof. Rømer's measurement of the velocity of light had assumed it to be a fact, but only with Bradley's discovery of aberration was there at last experimental evidence. Further observations showed Bradley an additional apparent motion of the star; this motion fluctuated over a nineteen-year period and was owing to a nodding, or nutation, of the earth's axis caused by the moon and its tilted orbit.

The aberration that Bradley measured for γ Draconis amounted to no more than 2″, though he claimed that his telescope was good enough for him to detect a change of 1″. Yet even with this accuracy he could detect no parallax, so it was obvious that even the nearest stars must lie at the most immense distances. Precisely how far depended on the size of the earth's orbit (because this was the baseline for his measurements) and determinations of this were difficult; we shall consider them in the next chapter, but can note here that in Bradley's time the generally accepted figure was 87 million miles. Because he had found that a star's parallax must be less than 1″, this meant that their distance was certainly more than 19 thousand billion miles—a fantastic figure.

One of the problems of measuring angles smaller than 1″ lay in the inferior nature of the telescopes astronomers such as Bradley had to use. Optically these were poor instruments, as they suffered from two defects or aberrations—spherical aberration and chromatic aberration. Spherical aberration was caused by the curved lenses, for these were made with their surfaces curved as part of a sphere, with the result that light from the edges of the telescope's front lens was not refracted to the same point or focus as light from the center. To the observer the result was that the center of an object appeared fuzzy when the outside appeared sharp. To overcome this, very long telescopes were made because the more gently the light was refracted, the less noticeable the fault; telescopes 150-feet or more long were made, though they were cumbersome things and one wonders how astronomers managed to use them at all.

Chromatic aberration was a more serious fault, especially when it came to making measurements. After all, a star only appears as a point of light in a telescope, and one star could be focused well enough, even if others nearby had to appear fuzzy; but chromatic aberration gave a colored fringe to every object, because light rays of different colors were each brought to a different focus. If one focused for red light, for instance, then the yellow, green, and blue images would be out of focus and give colored rings. The reason why the different colors that compose what we call white light were brought to a different focus by a lens was discovered by Isaac Newton. Each color was differently refreacted. As far as he was concerned, this meant that a refracting telescope with one front lens could never give a really successful image. What then could be done to improve matters? Newton developed a new design of telescope using no front lens at all (Chapter 7), but others decided there was another way out.

One way was to try to use a double lens at the front of the telescope: one lens of glass and a second lens made of something with a different power of refracting colors. Perhaps a hollow glass lens, filled with water, would do the trick? Newton tried this but without success, and others experimented with it too—notably the ingenious Robert Hooke. Not until 1729 was a practical solution found, and this was to use two glass lenses, one of crown glass (the glass usually used) and one of flint glass. Flint glass contains more lead and is denser than crown glass, and so refracts colors to a different degree; it was normally kept for making bottles and, because of its poor optical quality, had never been used for lenses before. The man who first hit on the idea of using flint glass was Chester More Hall, a lawyer who lived at Harlow in England and who was interested in the workings of the human eye. He realized that in the eye there was a lens and a watery substance, and the fact that one did not see colored fringes he put down to the combination of these two materials. His supposition was wrong, but his application of it to a telescope by using two lenses was successful. It did not remove chromatic aberration, but it did reduce it very considerably.

However, Chester More Hall did not publicize his success, and it might never have become well known were it not for a strange coincidence. Hall had not made the lenses himself but had commissioned two London opticians to construct them for him. He had used two opticians, one to fashion the flint lens and one the crown, in the belief that no one would know of his secret until he wished to make it public. But one of the opticians he employed did not carry out the work himself; he asked George Bass, the other one chosen by Hall, to do it for him, quite unaware that Hall had given Bass a second commission. The result was that one optician not only fashioned both lenses, but himself tried them in combination. He showed his result to John Dollond, a silk-weaver and amateur optician of great ability, who had also been trying combinations of lenses to overcome chromatic aberration. With the clue given by Hall's lenses, Dollond began to investigate the practicability of making nonchromatic, or achromatic, telescopes for sale, and in 1754 he and his son Peter were able to supply such instruments. Only telescopes with small lenses could be made, since flint glass presented all kinds of difficulties to the glass-maker, but if they were not good for exploring space, they were excellent for measuring positions and the Dollonds were soon supplying many for this purpose, including one to Greenwich Observatory.

The success that John and Peter Dollond achieved stimulated others to try to do better, but it was only in 1805, when the Swiss glass-maker Pierre Guinand had designed and developed a really efficient way of stirring molten glass, that larger lenses could be made from flint glass. By 1817 bigger refractors of high optical quality were being produced, particularly by Joseph von Fraunhofer in Germany, and with such instruments there was at last hope of measuring stellar parallax.

Success came in 1839 when Friedrich Bessel at Köningsberg Observatory, Germany, used one of Fraunhofer's telescopes on a special mounting that allowed the stars to be followed by one simple movement. For measuring positional changes Bessel had a new device, a heliometer. This had been devised by an Englishman, Servington Savery, almost a century before for measuring the diameter of the sun, and was really a front lens split in two across the middle. The two halves of the lens gave two images in the telescope's eyepiece, and as the halves slid over each other the two images could be made to join in the eyepiece. How much the halves had to be moved to achieve this was a measure of the separation of two objects (the opposite sides of the sun's disk, for instance) or of the change in position of a celestial body. Measuring by eye when two images touch each other gives more accurate results at night than the measurement of positions with a cross-thread and, in fact, Bessel reached the then astounding accuracy of being able to measure correct to $0''.33$.

The star whose distance Bessel measured is known as 61 Cygni —not a very bright object, but one that he believed was near because its observed movements in space were large. The actual parallax he obtained was $0''.35$. In the same year, the German astronomer Friedrich Struve using a telescope even larger than Bessel's announced that he had successfully measured the parallax of the bright star Vega. However the value Struve obtained $(0''.26)$ was three times too large according to modern measurements, and there is now considerable doubt about what Struve measured—certainly it could not have been parallax.

A year after Bessel announced his result Thomas Henderson, who had been appointed as Astronomer Royal for Scotland in 1834, gave details of the parallax of the star α (alpha) Centauri, whose distance he had measured in 1832–1833 at the Cape of Good Hope. The value he obtained for this star was the first genuine parallax measurement, but his natural caution had prevented him publishing his results until a colleague in South Africa had checked them by independent observations. By the time this had been done and Henderson was ready to tell the astronomical world it was January 1839; credit for the first successful measures had fallen to Bessel.

Bessel's measurements of 61 Cygni gave its distance as 600,000 times the distance earth to sun, but to put the value into miles and, indeed, to obtain a proper idea of the scale of the universe, it was necessary to have a really precise measure of the distance sun to earth. Because of its importance, this distance is known as the astronomical unit, and in the next chapter we shall examine the struggle involved in obtaining its value.

5 ✳ The Scale
of Space

It is no exaggeration to say that the astronomical unit is the most important single measurement in astronomy. Attempts to obtain its value were made as early as 280 B.C. when Aristarchos made some simple observations. His method was ingenious and is worth mentioning. His idea (Figure 5–1) was that if, at the moment

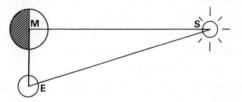

FIGURE 5–1. The method used by Aristarchos for measuring the distances of the sun and moon. The method was sound theoretically —measuring ∠ MES of the sun at the moment when the moon is at first quarter—but inaccurate because of the difficulty of telling when the moon is precisely half illuminated.

when the moon was at first quarter (that is, when it appears exactly cut in half), one observed the position of the sun (∠ MES), then the three angles of the triangle EMS could be obtained. These would allow a calculation to be made of the ratio of the moon's distance to the sun's distance; other observations should then allow one to obtain the moon's distance by itself. As we saw in Chapter 2, Aristarchos obtained a ratio, but his method was inaccurate because it is very difficult in practice to determine the precise moment when the moon reaches first quarter, and equally difficult to measure accurately ∠ MES. Nevertheless, it was a start. Aristarchos also believed that the distances could be found by making measurements of the shadow cast by the moon on the earth during a solar eclipse. This was somewhat complicated but worked well in theory, and about 150 B.C. it was actually used by Hipparchos. Hipparchos obtained an excellent value for the distance of the moon as 29.5 times the diameter of the earth (modern figure 30.1) and, coupling these results with observations of the length of the seasons, he calculated that the distance of the sun was 587 times the diameter of the earth. Again this gave the sun's distance as about twenty times greater than the moon's—a value that is some twenty times too small. In assessing the moon's distance, Hipparchos was therefore successful, but his measurement of the sun's distance was no better than Aristarchos'. His error was equally great, and this underlies the difficulty in measuring so vast a distance. The parallax of the sun is less than 9″ and, clearly, too small for detection by the observing techniques of the second century B.C. and, indeed, for detection by the naked eye. Only with the use of the telescope for measurement, and a proper understanding of the motion of the earth, could astronomers reach anything approaching a true figure, and after the attempts by Aristarchos and Hipparchos, the seventeenth-century French astronomers were the first to try to tackle the problem methodically.

The method used was based on Kepler's third law, the one connecting the orbiting periods of the planets and their distances from the sun. Obviously if the distance between the earth and another planet could be determined, as well as the orbital periods, then the distance earth to sun could readily be calculated. The advantage of this was that the distance planet to earth would be smaller provided the planets Venus or Mars were chosen, because both come closer to the earth on the heliocentric theory. However, observations when Venus is at its closest raise special problems, which we shall consider in a moment; the French were well aware of these, and, in consequence, they chose to make their first measurements on Mars at its close approach in 1672. Observations of its parallax were made from Paris by Jean Dominique Cassini, director of the observatory there, and in Cayenne on the northeast coast of South America by Jean Richer, one of Cassini's observatory staff. Their results gave a parallax of 9″.5 for the sun which meant that it lay 89 million miles away. This was the first figure that came close to the true value.

The success of the Mars measurements led astronomers to take up Edmond Halley's detailed proposals for determining the astronomical unit by observing a transit of Venus. When Venus is at its closest, it lies in between the earth and the sun and is invisible owing to the sun's glare, but once every 115 years or so Venus appears to cross, or transit, the sun's disk twice, the two transits occurring with an interval of eight years between them. On such rare occasions an observer on earth will see the planet appearing to touch the edge of the sun, travel across it, and then touch the other side. As Figure 5–2 indicates, if observations are made of

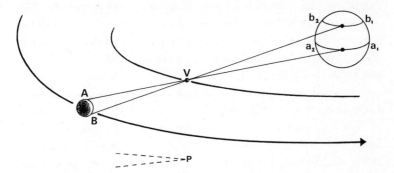

FIGURE 5–2. Halley's method for using the transit of Venus for measuring the distance of the sun.

the transit from two places widely spaced apart, then the observer at A, in the earth's northern hemisphere, will see Venus touching the sun at a_1 and a_2, while a second observer at B, in the southern hemisphere, will see Venus at b_1 and b_2. What Halley showed was that if the distance XY could be found, then the distance of the sun could be calculated because one already knew the distance AB, and also the ratio sun to Venus and sun to earth (from Kepler's third law). Halley made his proposals for observing and measuring a transit in 1716, but transits did not occur until 1761 and 1769.

In 1761 astronomers took advantage of the situation and sixty-two expeditions were dispatched to various places. In America

John Winthrop, professor of mathematics at Harvard, took a small party to Newfoundland and made observations there. By this time Halley's observing instructions had been slightly modified by the French astronomer Joseph Delisle, and all that observers had to do was to determine their longitude and latitude carefully and time the moment when Venus appeared to touch the edge of the sun. A second set of observations were made at the transit of 1769 by sixty-three expeditions. The observers found difficulty in determining the moment of contact of Venus with the sun's disk because of the optical illusion known as the black drop. It is caused by the glare of the sun's disk and the darkness of the disk of Venus—the blob of darkness seems to hold on for a few moments and thus makes it impossible to judge the precise fraction of a second at which contact is broken. Nevertheless, an exhaustive analysis of the observations were made by Johann Encke in 1824, which gave a figure of $8''.5$ for the sun's parallax and 95.25 million miles for the astronomical unit.

It is worth noting that, on his first voyage to the South Seas in 1768 to observe the 1769 transit, Captain Cook charted the whole of New Zealand and the eastern coast of Australia, annexing these lands as British possessions. Less important but sadly funny is a story attached to the earlier 1761 transit. The war then being waged between France and England resulted in a French astronomer, Guillaume Le Gentil, being quite unable to make his observations because he was prevented from landing. He decided to remain in the area and wait for the 1769 transit. The Académie des Sciences ordered him to land at Pondicherry in southeast India to observe it, not to go to Mauritius as he wanted. When the critical time arrived it was cloudy at Pondicherry and clear on Mauritius. By the time he returned to France in 1771, he had been presumed dead and his heirs were dividing up his estate; only after expensive legal battles did he overcome these tribulations! Fortunately no other observers suffered such privations and disappointment, and the success of the majority of the ventures could not be questioned. Fresh observations were made at the two transits of 1874 and 1882, and this time photography was used. We shall shortly come to the importance of photography for measurements of the astronomical unit—and for measurement of all kinds for that matter—but here it is enough to note that even with photography these observations were plagued by the black drop. All the same, computation gave a better value for the sun's distance—92.5 million miles.

Transits of Venus and the close approach of Mars were not the only means astronomers had for trying to obtain the vital measurement they wanted. The great French mathematical astronomer Pierre Laplace decided that he could compute the astronomical unit from observations of the gravitational effects of the earth on the moon and other planets. His calculations, made in 1787, eighteen years after the second of the transit of Venus observations, agreed well with the figure they obtained. But during the 1850's Laplace's value of 95 million miles was shown by Peter Hansen and Urbain Leverrier to be too large and, clearly, new measurements were required. Mars was observed in 1862 at a close approach and Simon Newcomb in Washington computed a value of 92.5 million miles.

All these variations meant that new ways must be tried so that confirmation could be given either to the higher values around 95 million or to the lower 92 million figure. In 1862, the French physicist Léon Foucault developed an instrument that had a rotating drum of mirrors, and with this he made a really successful measurement of the velocity of light. This, then, combined with a new and more accurate measure of Bradley's aberration of starlight (see Chapter 4) that Friedrich Struve had recently made, gave a value for the earth's velocity in its orbit. With this value, the distance of the sun could be computed using gravitation theory; the result was 92.25 million miles. The smaller value seemed, then, to be correct. But for so important and fundamental a measurement, stronger confirmation was sought, and still further attempts were made to measure the distance.

In 1887 the British astronomer David Gill, following a suggestion by George Airy who was then astronomer royal, observed a close approach of Mars. Gill made his observations from Ascension Island in the south Atlantic, obtaining the parallax of Mars from observations of the planet made in the early morning and in the evening, the earth's rotation displacing the position of his observatory sufficiently far to give him an accurate result. On calculating the sun's distance he found a value of 93 million miles; once again the smaller value seemed confirmed.

A new approach became possible after 1898, for then the asteroid Eros was discovered. The asteroids (now called plane-toids) were sought because in 1772 a strange relationship was found between the distances of the planets from the sun. If we write down the numbers 3, 6, 12, 24, 48, 96, and 192 [1] and then add 4 to each, and also start with 4, we have 4, 7, 10, 16, 28, 52, 100, and 196. Now, if we take the distance earth to sun to be 10 (it does not matter what the unit is; it could be ten times any-thing) then we find that 4 represents the distance sun to Mercury, 7 the distance sun to Venus, 16 the distance sun to Mars, 52 the distance sun to Jupiter, 100 sun to Saturn, and 196 sun to Uranus. The scheme breaks down for Neptune and Pluto, but in 1800 neither had been discovered, and the numbers appeared to tell something important about the solar system. This number clue was first discovered by Johann Titius who mentioned it as a foot-note to a German translation of a French book on science, but few took notice of it, and if it had not been for a young astronomer Johann Bode, it might have been completely forgotten. But Bode drew it to the attention of other astronomers, and it is usually known now as Bode's law.

If we examine the numbers we find, as Bode was quick to recognize, that there seems to be a gap. In the series, 10 is the distance sun to earth, 16 sun to Mars, and 52 sun to Jupiter. But what of 28? This represents the distance of no planet. Bode de-cided there must be an undiscovered planet lying between Mars and Jupiter, and when William Herschel discovered Uranus in 1781 and it was found that its distance also fitted into the series, considerable interest was aroused. The Hungarian astronomer Baron Xavier von Zach went so far as to compute "analogical" details of the orbit of the missing planet; this was in 1785 and he later began a telescope search for it. After fifteen years he still had had no success, but such was his belief that the planet

---

[1] We can work these out by taking 3, and then multiplying it by 2, then by 2 × 2, or 4, next by 2 × 2 × 2, or 8, then by 2 × 2 × 2 × 2, or 16, and so on.

existed that he managed in September 1800 to organize five German astronomers to take up the search: they became known by the wits of the day as the celestial police. While they were busy planning how they would make their search, and deciding who should scan which part of the sky, an Italian, Giuseppe Piazzi, who was making observations in Sicily so that he could prepare a new star catalog, noticed a spot of light that moved among the stars. He first saw it on January 1, 1801, and followed it for some nights before an illness interrupted his observations, but fortunately he wrote about his discovery to Bode. The posts were so slow that Bode did not receive the details until March 20.

In spite of the Renaissance and 200 years of scientific approach, there were still those who would have been more at home in a medieval setting. One of these was the German philosopher Georg Hegel, then a young man, who had just issued a pamphlet about the solar system. In this he proved conclusively from arguments, not from observation, that seven was the greatest possible number of planets. A presumptuous, totally unscientific document, it was scornful of Von Zach and his colleagues and, as we shall see in Chapter 10, was not the only document of its kind published about this time. Clearly, the importance of observation was still not fully realized, and some unscientific men felt they could make pompous pronouncements about the universe. But Bode was not to be deflected from his aim, and Piazzi's observations seemed to him the confirmation he had needed to find a planet to take up the empty space in his law.

The news spread rapidly and excitement grew because the body Piazzi observed had moved so close to the sun that it was lost in the glare, and there was doubt about whether it would ever be seen again. Piazzi's few observations were not enough to allow a good orbit to be calculated, but a young German mathematician, Karl Gauss, managed to compute one of a kind. By November 1801 he was able to announce an orbit, but cloudy weather prevented Von Zach's team looking for the body. Only on the last day of 1801 was Von Zach able to pick up the object, and his observation was confirmed the next night by Heinrich Olbers, a physician and amateur astronomer. At Piazzi's request, the planet was named Ceres after the goddess of his native Sicily.

Olbers continued to observe Ceres and soon found another similar object. This turned out to be a planet in an orbit very like that of Ceres, also lying between Mars and Jupiter. It was named Pallas. As the years passed a few more planets, all tiny like Ceres and Pallas—Ceres has a diameter of only 480 miles—were discovered and as of 1970 more than 3,000 planetoids have been recorded and orbits computed for more than 1,650 of them. Out of this vast number one, Eros, is particularly important for determining the astronomical unit. This is because it has a very oval orbit for a planet and, at its closest comes within 14 million miles of earth, nearly twice as close as the nearest point in Venus' orbit. In 1901 Eros came within only 30 million miles, but its small star-like appearance made it, rather than Venus or Mars, the planet used to determine the astronomical unit, because its parallax could be determined without the errors caused when observing a disk and allowing for its diameter. Observations were made internationally and were analyzed by Arthur Hinks, an astronomer in Cambridge, England. The analysis took a long time—there were no computers then—but in 1910 Hinks announced that the sun's distance was 92.83 million miles, and he believed that this figure was correct to within 62,600 miles.

In 1931 Eros made an even closer approach; this time it was only 16 million miles away. An international observing program was arranged under Harold Spencer Jones, later to become astronomer royal, but again it took almost ten years for the results to be analyzed and a value obtained for the sun's distance. In 1941, Spencer Jones was able to state that the sun's distance was 93.005 million miles and that this figure was correct to within 10,000 miles—that is, an error of no more than one mile in every 9,000.

One of the important discoveries that increased the accuracy with which observers in 1901 and 1931 could measure the sun's distance was photography. This had been developing for many years before 1901, but it was cumbersome and could only be used in bright light. The moon and the stars could not be photographed and, when this did become possible, it was some time before the results were good enough for precision measurement. How insensitive early photographs were can be judged from the fact that the first photograph of an outdoor scene taken in sunlight needed to be exposed for eight hours! This was in 1826, but eleven years later, after much research by two Frenchmen, Nicéphore Niépce, who had taken the photograph, and his colleague Louis Daguerre, a remarkable reduction in exposure time to between four and forty minutes was achieved. This was primarily due to Daguerre, and his photographs, which were taken on copperplates, gave some very beautiful and detailed results. However, to take a Daguerreotype was an involved process, because the photographer had to make his copperplate sensitive to light just before he took the picture, and then chemically treat the plate immediately afterward. Yet the American doctor John Draper managed to get a vague picture of the moon in 1840, and by 1850, William Bond in Cambridge, Massachusetts, had taken some really good photographs of the moon's disk.

The trouble with Daguerre's method was not only that it was cumbersome and its exposures inconveniently long, but also that there was no convenient way to make copies. A new approach, using sheets of paper instead of metal plates, was tried in England by Fox Talbot, an artist of little talent but a clever amateur scientist and experimenter. In 1839, helped by the astronomer John Herschel, who invented the terms negative and positive, Fox Talbot came on one of the most important factors that has led to modern photography—the development of a latent image. What this means in practical terms was that it became unnecessary to have exposures long enough to cause the photograph to blacken as was the case with a Daguerreotype; a short exposure followed by a chemical development made the picture appear. It was now possible to photograph objects as dim as stars. But there was one serious trouble with the method, and this was that paper was an unsatisfactory material for a negative, the paper fibers preventing a print with fine detail, and experiments were therefore made to find how to apply Fox Talbot's new light-sensitive surface to a glass plate. In 1847 Frederick Scott Archer managed to do this and also to obtain still shorter exposures; the astronomical uses of photography began to grow. Admittedly, the plates had to be made light sensitive just before use and exposed while they were still wet, but in England Warren de la Rue managed to take excellent pictures of the moon, and also designed a special photographic telescope—the photoheliograph—for taking daily photographs of the sun's disk; this was soon adopted in a number of observatories, and photography became established as a new and valuable way of observing.

But what of measurement? What was really needed was a simpler photographic process where one did not have to make the plate light sensitive just before use, and leave the telescope immediately after exposure. More research was needed, and eight years after Scott Archer's discovery, short exposure plates that could be used dry and developed at leisure were made by J. M. Taupendt and quickly improved by Richard Maddox and Charles Bennet who used gelatine, as we do today, to carry the new light-sensitive material. As a result, plates could be manufactured to a set pattern of sensitivity and a series could be processed together so that all received the same handling under the same conditions.

For precision measurement, the method adopted was—and still is—to take a photograph, and then to examine the developed plate through a measuring microscope. This, in spite of its complicated appearance, is really no more than a microscope fixed so that the photographic plate exposed in a telescope can be place underneath it and then moved from left to right and up and down. In this way the whole plate can be scanned and the position of any celestial body determined from the position of the plate. This technique has made it possible, for instance, for an astronomer such as Adrien van Maanen to measure positions of stars to $0''.001$—the equivalent of a human hair at a distance of ten miles!

But measurement in astronomy does not depend only on where something is in the sky, but also on when it is there. This lies behind all determinations of planetary position, of parallax, and of the astronomical unit, so there is no purpose in finding how to measure position correct to within $0''.001$ unless one can measure time correctly to a fraction of a second too. The development of the clock is a long and fascinating study that we cannot go into in detail and, though the Greeks made some time determinations using sundials and water clocks, it was not until the invention of the pendulum clock in 1673 that time-keeping was possible with any real degree of accuracy. Errors were still as much as a couple of minutes per day, but this was gradually improved. Instrument-makers seem to have become clear in their own minds about what was wrong with previous clocks and what was needed to put them right.

The basic principles behind an accurate mechanical clock are the regulator and the escapement. After all, a clock is merely a machine that is wound up, either by raising a weight or tightening a spiral spring, and is then allowed to unwind. To keep time it must unwind at a slow steady rate, and this is why it requires a regulator and an escapement; a regulator to regulate the rate at which the weight or spring may escape into the unwound state. It was Galileo who, during a service in the cathedral at Pisa in Italy, timed the swinging of a heavy lamp against the beating of his pulse, and discovered that a pendulum always takes the same time over each swing. And this is true whether the swing is large, as it is when we start something swinging, or small, as happens when the swinging is nearly completed. The total time it takes to go to and fro is always the same for a pendulum of any given length. From this it becomes obvious, as Galileo realized, that a pendulum made of something heavy swinging on a rope, a chain, or a pivoted rod was the perfect regulator for a clock.

Two of the main problems with a pendulum were, first, to keep it moving and, second, to allow it to regulate the clock without interfering too much with its freedom to swing. (See Figure 5–3.) Strike a pendulum one way while it is moving in another and you upset the equal swings in equal times property; make the pendulum drive clock gears so that it knocks gear teeth at moments when it is moving most quickly—at the midpoint of its swing— and again you interfere with its timekeeping ability. Clockmakers therefore set about designing escapements that interfered with the pendulum's swing as little as possible, and never at the midpoint, and so accuracy was improved. A dead-beat escapement—one that allows the pendulum to beat time and regulate the clock with as little interference as possible—was invented by George Graham in 1715. Its success rested only on details such as the shape of gear teeth and the two teeth from the pendulum that touched them, but with it the error in timekeeping could be reduced considerably.

A third major problem was caused by the way a pendulum

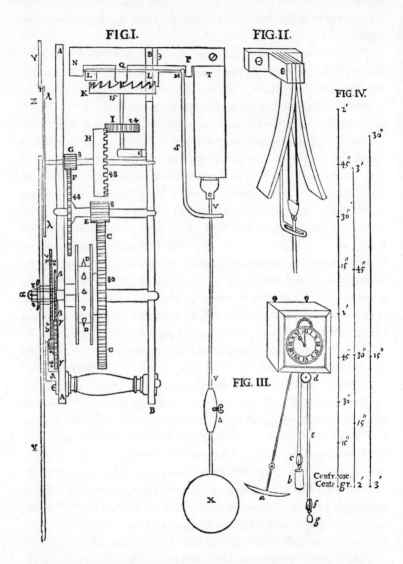

FIGURE 5–3. Pendulum clocks designed by Christian Huygens on the basis of his theoretical work on pendulums. Such clocks gave a new standard of accuracy and, with modifications, did duty for astronomers well into the present century. From C. Huygens, *Horologium oscillatorium* (Paris, 1673). Courtesy of the Ronan Picture Library and E. P. Goldschmidt & Co., Ltd.

changed its length with fluctuations in the surrounding temperature. Graham got over this by using a pendulum made up of separate brass and iron rods, because temperature affects some metals more than others. The rods were of different lengths and, as the temperature increased, they expanded, the brass expanding more than the iron; they were arranged so that whatever the expansion, the distance between the top of the pendulum where it was pivoted, and the bob at the bottom, remained the same. Because a pendulum of given length always swings at the same rate, Graham's gridiron pendulum kept time very well. This, coupled with his dead-beat escapement meant that his clocks kept time to within one second every twenty-four-hour-day.

Things had now reached the stage where the way to tackle the two most important parts of the clock mechanism had been amply demonstrated. As in so many cases of this kind, it did not fall to the innovator to carry his own inventions to their ultimate peak of development; Graham's protégé, John Harrison, built the first clocks accurate enough to have errors of only a fraction of a second. In 1759 he built an instrument not much larger than a large pocketwatch, with an average error of no more than about .05 second per twenty-four hours. This was quite remarkable and solved the problem facing sailors who needed to find their longitude at sea (see Chapter 6).

But astronomers wanted greater precision still, and in the 1840's Sigmund Riefler, a Munich instrument-maker, built some observatory clocks in which every care was taken to give the pendulum its impulse at just the right moment and to keep it in a case so that the air pressure and temperature changed scarcely at all. These clocks kept time correct to within .01 second per day, but astronomers were not to remain satisfied for long. By the early twentieth century, scientists working on electrical and radio research found themselves dealing with impulses that changed thousands of times a second, and in order to measure these they required clocks with an error of no more than .003 second per twenty-four hours. This incredible accuracy was achieved in 1924 by the engineer W. H. Shortt, who ingeniously used two pendulums to drive a single clock. One, the master, was kept at a constant temperature and pressure and did no work in driving the clock, but merely received a push when necessary from the second pendulum. This, the slave pendulum, drove the clock and received pushes from the clock mechanism. The pendulums were connected electrically and the master kept the slave in time by feeding back part of the impulse received from the slave. The timekeeping of these free pendulum clocks was remarkable, for they had an error of about only .0025 second per day.

Since 1924 demand for still greater precision has increased, and it soon became clear that even electrical/mechanical clocks, such as the free pendulum, could not provide it. Designers had to turn to purely electrical methods, and during the 1930's the quartz clock was invented. Quartz, a mineral that occurs naturally in the earth's crust, exhibits a piezoelectric, or pressure-electric, effect. If it receives impulses of electrical current, it acts as if it were being pressed and vibrates. In a quartz clock, a ring of quartz is made to vibrate by feeding it with electrical impulses. The ring has its own natural speed or frequency of vibration, just as a pendulum does, and the electrical impulses set it vibrating at this frequency—the quartz is the clock's regulator. But while the quartz vibrates, it sends out electric impulses itself, and these are used to correct both the impulses coming in to it, and to drive the clock. Held at constant temperature and pressure, the quartz clock can keep time to an accuracy of a few thousandths of a second per twenty-four hours, a degree of precision that amazed astronomers at Greenwich and Edinburgh observatories when the clocks were installed in 1935. Things have not rested there, and in 1955 a successful atomic clock was built at the National Physical Laboratory in England. In this, instead of quartz, vibrating gas atoms are used, which are more than thirty times as accurate as the quartz clock. Indeed modern timekeeping is so precise that whereas astronomers used to check their pendulum clocks by the rotation of the earth, using a transit circle, they now use an automatic zenith tube, developed from Graham's design made for Bradley and check the rotation of the earth from their atomic clocks! (See Figure 5–4.)

Such phenomenal timekeeping is necessary in these days of space shots and the increased use of ultra-high frequency radio waves in radio astronomy, in radar, and similar developments. It has allowed the astronomical unit to be determined even more precisely, for radar pulses (see Chapter 12) have been bounced off Venus. Using the results of these, and also the results of photographic measures of the distance computed from observations of Eros made between 1926 and 1945, Eugene Rabe of Cincinnati Observatory has worked out that the astronomical unit is 92.957 million miles.

FIGURE 5–4. An atomic clock, the most precise of all clocks for measuring short intervals of time. Courtesy of National Bureau of Standards, Boulder, Colorado.

6 ✳ Organization of the Stars

The earliest watchers of the skies organized the stars into groups, imagining that they depicted the characters and beasts that populated their folklore, and the animals they saw around them. This was only the first stage, for as soon as their observations became more accurate, it was necessary to be more precise; astronomers had to move from the picturebook stage and begin to express star positions in numbers. At least this was so in theory, though stars were often referred to in both ways: first, descriptively—in the left shoulder of Orion, for example—and then, by numbers that denoted the celestial longitude and latitude of each, that is, its position along the ecliptic and above or below it. In this way the first star catalogs were compiled, but some people thought it irreverent to reduce celestial bodies, who might well be divine beings, to mere numbers in a list, and in 150 B.C. Hipparchos was accused of impiety when he drew up his catalog of 850 stars.

The position of stars was not the only information cataloged; there was also the question of brightness. The brighter a star seems to be, the larger it looks, and by the time of Hipparchos magnitude was used as the technical term for brightness. Hipparchos divided the stars into six classes, the stars of the first magnitude being the brightest and those of the sixth magnitude, the dimmest, and this procedure was followed by later astronomers. Catalogs were prepared by Aristyllos and Timocharis about 280 B.C., by Hipparchos, by Ptolemy in 150 A.D., by the Muslim astronomer Ulugh Beg, and by Tycho Brahe, who drew up what we may call the first modern catalog. Tycho accurately measured his 777 star positions using what he believed to be more convenient reference points—right ascension, along the celestial equator, and declination for north and south of it. For more than a century after he issued his catalog in 1598, it was a standard reference work. When Johann Bayer returned to the picturebook idea and published his star maps illustrating the constellations, he based his atlas on Tycho's catalog and added Greek letters to the brighter stars, thus introducing a useful shorthand that astronomers have followed ever since.

It would be tedious to even mention all the other star catalogs and atlases that were prepared, but, of course, as observing became more precise, so the catalogs and charts became more accurate. The stars of the southern skies were cataloged, especially by Edmond Halley and John Herschel, and ordinary letters, code letters, and numbers were used to refer to particular stars because the number observed and needing mention outnumbered the letters of the Greek alphabet. In addition, star names were often taken over from the Muslim astronomers, which is why we have what at first sight seem strange names, such as Betelgeuse, Aldebaran, and Caph. But it is the results of the cataloging that are important to us here, for there are three discoveries that arose from the preparation of catalogs: the precession of the equinoxes, stellar motion, and the scale of brightness.

Hipparchos discovered that all the stars seemed to change position in a way that can best be described by saying that the celestial equator appears to move in the sky. This movement causes the points where it cuts the ecliptic to move slowly backward across the heavens. When the sun is at the points on the ecliptic that cut the celestial equator, the day and night are of equal length—it is an equinox—so the backward movement became known as the precession of the equinoxes. Hipparchos found this by taking his own observations and those of Aristyllus and Timocharis made a century and a half earlier; the star positions had altered by an amount that was too large to put down to mistakes in observing. The change was 2° over 150 years, and so Hipparchos worked out that the change was 48″ per year. Later on astronomers determined precession with greater accuracy, and today we accept a figure of 50″.25. For all other studies of star positions this figure is of great importance. The cause of precession remained a mystery until the time of Newton and his theory of gravitation. Only in 1687 with the publication of the *Principia* was it shown to result from a motion of the earth's axis. Because the earth is not a true sphere but has a bulge at the equator and because its axis is tilted over at an angle of 23°.5 to its orbital path around the sun, the sun's pull on the earth is not straightforward; it produces a twist. The twist causes the earth's axis to move so that each end describes a circle (Figure 6–1). The moon also has a similar effect.

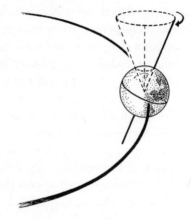

FIGURE 6–1. The changing tilt of the earth's axis known as precession. It gives observers different pole stars as the years go by.

As a result the earth's equator moves, and this in turn moves the celestial equator to give precession. The movement of the earth's axis is slow, and it takes 26,000 years to complete one revolution. For the visual observer, this means that he has a different pole star as the centuries pass—for example, in 3000 B.C. the star α (alpha) Draconis was the pole star, and in 7000 A.D. it will be α Cephei. Though the pole star might change, the constellations remained fixed, and it was not until the eighteenth century that the question of whether or not stars actually moved arose.

This question was raised when Edmond Halley compared his own and other contemporary cataloged star positions with those of Greek catalogers, especially with the catalogs of Timocharis, Hipparchos, and Ptolemy. His aim in doing so was to check the value for precession, but he also found that three stars contradicted the value he obtained from all the others. These three were Aldebaran, Arcturus, and Sirius, all bright and easily seen. But could the apparent shift in position of these three stars be owing to errors in observation? In view of the correctness of the rest of the cataloging, observational error seemed out of the question, especially because all three ancient observers agreed in the figures they gave. So in 1718 Halley concluded:

> these Stars being the most conspicuous in Heaven, are in all probability the nearest to the Earth, and if they have any particular Motion of their own, it is most likely to be perceived in them, which in so long a time as 1800 Years may shew it-self by the alteration of their places, though it be utterly imperceptible in the space of a single Century of Years.

It was some time before the investigation of stellar motions was followed up, or even before Halley's conclusions were generally accepted, though he had demonstrated that Sirius showed a movement that could be detected from his own observations compared with those of Tycho made only a little more than a century before. The next step had to wait until Tobias Mayer began cataloging stars during the 1750's. Mayer measured some more stellar motions and suggested that if the sun itself were a star—and this idea was beginning to find favor—then it too should move. He made suggestions about how such a movement could be detected, but the man who made the first practical attempt to do this was William Herschel.

Born in Hanover in 1738 and trained as a musician, Herschel emigrated to England, which was then ruled by George II, himself a Hanoverian. He began to earn his living by teaching music, with astronomy as a hobby, but when in 1781 he discovered the planet Uranus his name was made. This discovery captured public imagination because all the other planets had been known since primitive times, and it was the first planetary body to be found in recorded history. Herschel came on the object quite by chance while he was making one of his detailed and laborious surveys of the sky. At first he thought it was a comet because it showed a disk and moved among the stars, and, if one happens to detect a comet with a telescope before it comes very close to the sun, it shows a disk but no tail. However, later observations and calculations by other astronomers demonstrated that the path was too circular for a comet. Once recognized as a planet, it had to be named; Herschel wanted to call it Georgium Sidus, or George's Star, after the king, but others decided it would be better to keep to classical tradition, and it was named Uranus. Nevertheless, Herschel received a royal pension, and became royal astronomer—but *not* astronomer royal, for Herschel's appointment was a far more personal one than that of the director of Greenwich Observatory —his one duty being to display the heavens through a telescope to the royal family. Herschel was a great observer and concentrated his attention almost entirely on the stars rather than the planets, using telescopes that he constructed himself. We shall come to these in the next chapter, but will look at some of his stellar work here.

By the 1780's the motions of fourteen stars had been determined, and Herschel decided to see whether he could discover the sun's motion from these. His argument was that if one looked into the sky in the direction of the sun's motion, one should observe the stars appearing to move away from this point. Conversely, if one looked in the opposite direction, the stars would appear to move toward one another. It was an argument based on the kind of everyday experience we have when we travel in an automobile. As we drive along a tree-lined road, the trees far in front seem to part and move sideways, to the left and right, and then they pass the sides of the automobile. If we look through the rear windows and follow them after they have passed, they seem to close together as the road runs away into the distance. It is all an effect of perspective, and though Herschel knew nothing of automobiles, he considered what would happen if one was on board a ship passing through a fleet of other ships.

After analyzing the motions of the fourteen stars whose movements had been measured, he discovered that the sun seemed to be moving toward a point in the constellation of Hercules. This was in 1783, but most astronomers refused to set much store by Herschel's result because, quite rightly, they were of the opinion that fourteen stars were too few to give a reliable answer for a study that was essentially a statistical one, for large numbers are required to give sound statistical results. Only after 1830, when

Friedrich Argelander at Bonn, Germany, obtained almost the same result from a study of the motions of 390 stars, did astronomers accept the fact of the sun's motion. That the sun moved just as the other stars do made it reasonably clear that it was no more than a star and only occupied so important a place in our scheme of things because we are comparatively close to it. This outlook had a great effect on the attitude astronomers took to their discoveries about the chemical nature of the stars (Chapter 11) and, together with the recognition of the motion of all other stars, was of importance to them as they tried to piece together the layout of stars in the universe (Chapter 8).

But Herschel's basic studies were not confined to the motion of the sun; they were mainly concerned with the stars themselves. He was the most hard-working stellar observer there has ever been, for he surveyed the whole of the sky visible from the latitude of London (51° 30′) no less than four times. He picked what he called 3,400 gage fields of stars and counted the number of stars in them, and made many discoveries of new objects (see Chapter 8). He was the founder of systematic photometry—the regular and systematic study of stellar brightness—and developed a method for precisely measuring a star's brightness. To do this he used two telescopes, one larger in aperture than the other. He then selected two stars so that the dimmer appeared as bright in the larger telescope as the brighter one did in the smaller; knowing the difference in size of aperture he was able to work out the difference in brightness.

Herschel's methods were improved by other astronomers, not least by his son John, who in 1834 began comparing a star's brightness with an artificial star lit by a lamp attached to the telescope. The comparison method using an artificial star—John Herschel called it an astrometer—remained in use until the widespread adoption of photography for stellar measurements. The brighter a star, the larger the blob it makes on a photographic plate, so the size of blob is a measure of brightness. Consequently, if a photographic plate is examined under a measuring microscope, brightness can be obtained from a survey of the diameter of the blobs. The amount of darkening of each blob is another indication of brightness, and in 1899 a special microscope-photom-

eter, or microphotometer, was designed by the German astronomer J. Hartmann. Though more sophisticated, the machine used
today still works on the same principle.

FIGURE 6-2. A photograph of the stars in the field of the constellation of Orion. The brighter the star, the larger the blob it gives
(the blobs do not depend on the star's real size). The bright star
known as Betelgeuse, in the top left-hand shoulder of Orion, cannot be seen because it is red and this photographic plate was only
sensitive to blue light. The grid of lines appears because this was
part of a photographic chart of the sky made by Franklin Adams
between 1902 and 1905. Courtesy of the Ronan Picture Library.

Brightness obtained photographically is not straightforward,
because the photographic plates used in astronomy are not equally
sensitive to light of all colors, being more sensitive to blue than
yellow and red. (See Figure 6-2.) In consequence astronomers

have had to take great care in assessing star colors and have devised a method for stating colors according to a number scale. The scale is derived from measures of brightness taken from some photographs with plates sensitive to blue only, and some in which only green, yellow, and red light is allowed to filter through. However, an investigation of this kind requires a knowledge of brightness that is detailed enough to be precisely expressed in numbers, and the first steps toward such a scale were suggested by John Herschel. He noticed that in the six magnitudes used by Hipparchos, the brightest stars were 100 times brighter than the dimmest. This was taken up and investigated in detail at Oxford by N. G. Pogson in 1850, who worked out a scale based on the brightness measurements that Argelander and Friedrich Bessel published in their great star catalog *Bonn durchmusterung* (*Bonn Review of the Stars*). He decided to call the dim stars in the catalog sixth magnitude, and then took the first magnitude to be 100 times brighter. This made each magnitude 2.512 times brighter or dimmer than the next magnitude, depending which way one reads the scale—magnitude 1 stars being 2.512 times brighter than stars of magnitude 2, and so on.[1] With such a scale very bright stars lay beyond the first magnitude, a star like Vega having a magnitude of 0 because it is 2.512 times brighter than a first magnitude star such as Spica, whereas a very bright star like Sirius has a magnitude of −1.4, which means that it is 2.4 magnitudes up the scale from Spica or, in other words, 9.1 times brighter. There is no limit to Pogson's scale, and on it the full moon has a magnitude of −11 and the sun of −26.7.

With ideas of brightness in mind, even if not mathematically precise ones such as Pogson's, astronomers were bound sooner or later to ask themselves whether stars stayed at one brightness all the time. There was, after all, some evidence that might lead one to suppose otherwise, Tycho's supernova of 1572 being a case in point. But to turn to something less spectacular, in 1596 the astronomer David Fabricius noticed a star in the Cetus constella-

[1] The unusual figure of 2.512 is used because there are five gaps going from magnitude 1 through 6, and 2.512 multiplied by itself 5 times gives 100.

tion fade until it was no longer visible, and he thought that he was observing a nova sinking back into insignificance. Yet thirty-five years later the star was seen again by Guillaume Schickhard who noticed it was as bright as Mekab, the brightest star in the constellation, and in 1639 the Dutch astronomer Phocylides Holwarda noticed it drop in brightness again, disappear, and reappear. He found that this variation took about eleven months but that the amount of change was irregular. The star was given the name Mira (miraculous one), and there is some evidence that its variation had been noticed by the Babylonians.

In the years that followed, variations were discovered in other stars, notably Algol (in Perseus) and stars in Taurus, Serpens, Cygnus, Canis Major, Leo, Sagittarius, Argo, and Lupus. But were these all there were? How many stars, in fact, varied their light periodically? To answer such questions meant undertaking a systematic search of the same areas of sky time after time not only to detect stars that varied, but also to find how they varied with time. In 1783 John Goodricke, a nineteen-year-old amateur astronomer who was both deaf and dumb, observed Algol, noted its light variation, and suggested that its variation was caused by a second star orbiting round a brighter companion. This was not the first suggestion of the existence of a double-star system, but it was the first time the idea had been proposed for scientific reasons, and we shall return to it shortly. Unfortunately Goodricke died when he was twenty-one, and though he might have paid some attention to other variable stars had he lived longer, the first to catalog them carefully was William Herschel.

Herschel cataloged and also measured the variation of the small number of variable stars known by the second half of the eighteenth century. He then set about what he called a method of sequences—observing an area of stars, noting their brightness, and then coming back at a later date to the same area and noting brightness again. The method was tedious but he did discover the variable star Ras Algethi (α Herculis), which had, he estimated, a period of variation of sixty days and so lay between Algol with its period of almost three days and Mira, which varied with a total period of eleven months.

Argelander and John Herschel were the next major contributors to our knowledge of variable stars. Argelander appealed to amateur astronomers to help because the study of variables was ideally suited to their smaller telescopes, and in the mid-nineteenth century a great effort was made to observe and catalog them. By 1884 John Gore in Ireland was able to publish a list of 190 variables and, four years later, to add another fifty-three. The American S. C. Chandler also published a catalog of 225 variables in 1889, and in 1919 Gustav Müller and Ernst Hartwig were able to provide astronomers with a really definitive catalog listing more than 900 variables, sorted out according to how long they took to complete their variations. This great increase in numbers was owing to new observing methods, especially the development of the stereocomparator and of a light-sensitive photoelectric cell.

Strangely enough, the first of these was a late nineteenth-century development that arose from popular interest in taking pairs of photographs to give a stereoscopic or three-dimensional effect. Everyday things were photographed first of all, but then pictures were taken of the sun and moon. The stereocomparator was a device for using two photographic plates of the stars and rapidly switching vision between one plate and the other. If the two plates were the same, then no change could be seen, but if there was any difference—because of the movement of one star among the others, say—then the image of that star would seem to jump or blink. The device gradually became called the blink microscope and was extensively used for picking out stars with detectable motions and variable stars. Only since World War II have electronic comparators superseded the blink microscope.

As far as the photoelectric cell is concerned, this arose owing to some work by Joel Stebbins at Mount Wilson Observatory during the first decade of the twentieth century. It was developed in 1911 by two German physicists, Julius Elster and Hans Geitel, and was almost immediately applied at Berlin and Tübingen observatories, giving an increase in accuracy of magnitude determination. The device has since been widely used by professional astronomers, and in 1937 and 1938 Gerald Kron at Lick Observatory further extended its sensitivity so that he could, for example, determine the changes of brightness of Algol with an error of no more than .002 of a magnitude.

One of the aims in measuring brightness was to find out why some stars behaved in this way. Goodricke's suggestion that Algol varied because of a dimmer star orbiting about it was one explanation, and it might fit more stars than Algol. It seemed a highly probable reason in the light of William Herschel's investigations in another direction—the study of double stars. Here Herschel was really trying to determine stellar parallax (Chapter 5), and he thought he would have the best chance of detecting a shift if he carefully observed pairs of stars close together in the sky. He based his argument on the idea that stars only appeared close together by chance and because they happened to lie nearly in the same line of sight. He imagined that they were really immensely far apart and did not believe they were physically connected in any way. In 1767 a rather obscure astronomer, John Michell, had suggested that there was some actual connection, and when seventeen years later Herschel made it known that his observations showed that a considerable number of pairs existed, Michell once again attracted attention to his ideas.

Before Michell's second attempt to get his views accepted, Herschel had said he thought that it was "much too soon to form any theories of small stars revolving round large ones," but twenty years later, in 1802, he had changed his mind, and by 1804 was in a position to lay before the Royal Society proof that he had detected fifty orbiting pairs of stars. Others were not long in following, using refined measuring techniques, first at the telescope eyepiece with micrometers and then by photography. By 1837 Friedrich Struve had enough evidence to publish a catalog of 3,000 doubles, and in subsequent years their positions were measured correct to $0''.01$. This work was followed by Struve's son Otto, and in America by Sherburne Burnham at the Lick and Yerkes observatories. The Struves and Burnham made their measurements by optical observation, between 1914 and 1917 Ejnar Hertzsprung did a great deal of photographic work at Potsdam. Now, after 130 years, enough is known to make it look as if half the stars we see are binaries or still more complex systems comprising three or even four stars in mutual orbits around one another.

The discovery of so many binary stars has shown that only a

very few vary their light because one member of the pair passes in front of the other as they orbit; most do not vary their light at all. Goodricke was right about Algol showing variation owing to this cause, but it has turned out that Algol is an exception: most variables change brightness for other reasons (Chapter 8). But binary stars are important in their own right because, by observing their mutual orbits, it is possible to use gravitation theory to compute how massive these stars are. Indeed, binaries provide the astronomer with the only way of determining stellar masses.

At the large university and particularly national observatories, research into the nature of stars and the kind of universe that we find exists in space has come to occupy the major part of their programs. But this is not why national observatories were originally founded, and during the seventeenth century the observatories at Paris and Greenwich were established solely for the purpose of determining the positions of celestial bodies. The reason behind this was that such measurements seemed to offer the only practical solution to the vital problem that then faced such maritime nations as England and France—the determination of latitude and longitude at sea.

Finding latitude presented very few difficulties, because a mariner had only to determine the elevation of the pole star (or more precisely the celestial pole) above his horizon. If the pole star was obscured, he had only to observe the altitude above his horizon of some other known star, or if he wished to determine latitude by day, he could observe the sun's altitude. Comparing these altitudes with the declination above or below the celestial equator given in the star catalogs, he could quickly find his latitude. No elaborate instruments were necessary and the astrolabe was used for this kind of determination until improved devices were adopted during the sixteenth century. But the problem of determining longitude is another matter, as Figure 6–3 illustrates. Suppose we are in a ship A and we wish to find our longitude, that is the angular distance ANS. The only way to do this is to find noon at A and at S. If the sun is directly overhead at A (noon at A), we need to find out how long before it was overhead (noon) at S. If it is local noon at S four hours before it is noon at A, then we know that the sun has moved 4/24 or 1/6 of its apparent journey round the earth. We then know that S is 1/6 of the way round the earth from A and, measuring longitude in degrees, we can see that S is west of A by 1/6 × 360, or 60°.

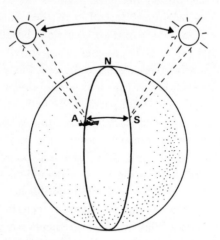

FIGURE 6–3. The problem of finding longitude at sea, which really means finding the time when a ship is at A, compared with the time at a standard observatory S.

The trouble in the sixteenth and seventeenth centuries was that there were no reliable timekeepers that could be taken on a voyage. The best a mariner could do was to use dead reckoning, which meant calculating from the speed of his ship and his compass readings how far he had traveled and in what direction. But dead reckoning was not good enough; it was all right when a ship hugged the coast, but once it ventured into the open sea out of sight of land, its captain could easily lose his bearings. There are many stories of ships going round in circles, and others touching land 500 miles away from the port they were aiming for, and by the seventeenth century the problem became so serious that governments intervened. In 1664 Jean Picard in Paris caused a stir by pointing out that in France there were no accurate standard instruments for determining latitude, leave alone longitude. Drastic action was needed to put matters right, for if mariners had no standards from which to begin, they were in an extremely dangerous position. Louis XIV's attention was drawn to the situation by his ministers, and he immediately established an observatory at Paris, where Picard came to work, and Jean Cassini was appointed the first director. Within fifteen years Cassini was publishing an annual list of dates, times and positions of celestial objects; the *Connaissance des temps;* it is still published today.

In England, Charles II appointed a committee to consider longitude, and in particular to discuss some practical suggestions that had been made for finding longitude at sea from observations of the moon. The basis of the idea was that inasmuch as the moon moved across the background of the starry sky, the mariner could treat it like the hand of a clock and, by measuring the angle between its edge and a nearby star, obtain the standard time (at A). Once standard time had been determined, local time was not hard to find. But to use the moon as a clock hand meant, of course, having tables of the moon's position to consult. Owing primarily to the advice of John Flamsteed, the committee reported to the king that, though admirable in theory, the method was unworkable in practice as the moon's position was not known with sufficient accuracy. The king's answer came in 1675 when he established an observatory "within our park at Greenwich, upon the highest ground . . ." and appointed Flamsteed, who seemed so well informed, as his astronomical observator, or, as he later came to be known, astronomer royal.

Both Paris and Greenwich were therefore institutions founded for the solution of a particular problem, and both have been famous for the great accuracy of their measurements. As mentioned in Chapter 5, the difficulties of determining longitude at sea were solved in 1759 when John Harrison constructed a clock that really could keep accurate time at sea in all weathers; but clocks can go wrong, they may get broken, and so it is helpful to have a second way for the mariner to find his longitude. Consequently, Harrison's invention, and the subsequent improvements made by a succession of fine clockmakers, did not render astronomical observations obsolete. The method of lunars was still developed by Flamsteed's successors, and in 1763 Nevil Maskelyne, an independent astronomer, began publishing astronomical tables for navigators under the title of the *British Mariner's Guide.* After he had been appointed astronomer royal in 1765,

FIGURE 6–4. A meridian circle in use at the Paris Observatory during the 1880's. The telescope is set accurately in a north-south direction, and the moment of meridian passage of stars is precisely determined. (Compare with Figure 4–1.) This gives the positions along the celestial equator, known as right ascension. The altitude at which the telescope points is read off through microscopes from the circle to the left; this gives the star's declination above or below the celestial equator. From C. Flammarion, *Astronomie Populaire* (Paris, 1881). Courtesy of the Ronan Picture Library.

Maskelyne revised the publication, partly with new material and partly with the preparation of excellent lunar tables by Tobias Mayer, and in 1767 the *Guide* was superseded by an official *Nautical Almanac*, which is still published each year.

Greenwich and Paris were the first purely astronomical institutions to be established in the Western world, and they are the direct progenitors of all later astronomical observations. Through their original navigational purpose has gone, the fact that their directors were not narrow-minded computers but men of broad vision who saw the importance of astronomical research has meant that they are still flourishing organizations that, by their example, have led to the establishment of other observatories all over the world.

7 ✳ The Big Telescope

One of the most important aids to the modern astronomer is the large telescope. It has allowed him to look far into the depths of space and to photograph objects that cannot be observed with smaller instruments. The reason for this is the vastly increased sensitivity of a telescope with a large aperture; for instance, the 100-inch telescope on Mount Wilson can observe objects 2,500 times dimmer than a telescope with a ten-inch aperture, and the 200-inch can detect objects 10,000 times dimmer than is possible with a ten-inch. There are technical problems that arise as soon as we try to increase aperture, and it was a long time before really big telescopes could be made. When Wilhelm Struve used a fifteen-inch refractor in 1838, astronomers envied him because of the telescope's size and optical excellence, but sixty years later, when the world's largest refractor began work in Yerkes Observatory, the aperture had only crept up to forty inches. For probing the depths of space this was not enough, yet it seemed to be the practical limit for the refractor. For one thing, its large double lens had to be so thick to prevent it sagging in the middle that it absorbed too high a proportion of the incoming light. But there were other difficulties. It is extremely difficult to construct a large lens free from bubbles and strains, and even if this problem is overcome, there is still the question of focal length. A refractor must have a long focal length in order to minimize its optical defects, and this means a long tube; in the case of the Yerkes telescope it had to be sixty-three feet. But such a long tube must not be allowed to bend, even by a fraction of an inch, and this is very hard to prevent. So after the Yerkes telescope, the world's largest instruments were designed as reflectors because they can be shorter and present less difficult optical problems. In fact, large reflectors had been made long before the 1890's, but they too had their disadvantages, and all along deciding which to use was a matter of balancing the qualities and defects of each type. The refractor was preferred for measurement, but the reflector was adopted when probing space was the prime consideration.

The construction of the forty-inch Yerkes refractor was an immense achievement, but it seems to represent the peak of the refractor's development. Fortunately a little before the refractor was built, a new technique was invented for making mirrors and, coupled with the growth of heavy engineering, meant that the reflector could really come into its own. But the story of the reflector goes back to the middle of the seventeenth century though, as with the refractor, there is uncertainty about its origin. The first definite information about it was published by the Scottish mathematician James Gregory in 1663, in his *Optica promota* (*The Advancement of Optics*). Gregory used a concave mirror instead of a lens to collect the light and bring it to a focus, but he came up against the difficulty that faces everyone who wants to make a reflector rather than a refractor—how to arrange things so that an observer can use the telescope. The problem is that in a reflecting telescope, the light passes down the tube and is reflected back again by the mirror, so that to observe the result one ought to place one's head in front of the mirror, but this is no use as the incoming light would be blocked out. Gregory's solution was ingenious. He suggested placing a small concave mirror in front of the main mirror (Figure 7–1a) and reflecting the light back again, down the tube, allowing it to pass through a hole in the center of the main mirror. The hole did not matter because the front mirror stopped light from reaching the middle of the main mirror anyway. But this was all in theory; he never made it himself.

Isaac Newton severely criticized Gregory's design because a second concave mirror would give the telescope a very small field of view, so that only a fraction of the sky could be observed at any time. It was in fact the reflecting equivalent of Galileo's telescope, which had been cast aside in favor of Kepler's design with a wider field of view. One further disadvantage was that Gregory's method would introduce all kinds of optical errors, or so Newton claimed. As there was no available alternative to Gregory's reflector, Newton decided to provide one himself and not only drew up plans, but actually constructed an instrument to prove his point.

Newton's urge to make a successful reflector was based on his mistaken belief that the refractor could never be cured of chromatic aberration (see Chapter 4). His opinion was a result of experiments he made when at home during the plague years 1665 and 1666. He had then worked on the whole question of refraction and the nature of colored light, and in a famous series of experiments using prisms, he conclusively proved that white light is, in reality, a mixture of light of every color from red at one end of the range or spectrum, to violet at the other. He showed that each color is reflected by a different amount, red being refracted least, and blue most strongly.

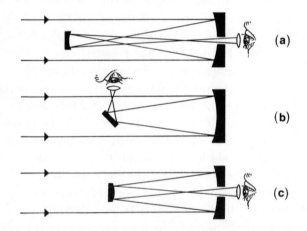

FIGURE 7–1. The three main kinds of reflecting telescope: (a) Gregory's design; (b) Newton's arrangement, and (c) Cassegrain's system.

Newton wrongly concluded that glass of different kinds behaved in so similar a way that he finally decided that the reflecting telescope offered the only solution. To get as wide a field of view as possible, he took Kepler's basic design of the telescope, but bent the light beam (Figure 7–1b). He turned it to the side of the tube and, in his earliest drawings, adopted a small flat mirror to move the light through a right angle. This made boring a hole in the main mirror unnecessary, but because the observer did not peer directly through the tube, sights had to be fitted so that the telescope could be lined up on a celestial object. Newton already had some practical experience in optics, for when he was home at Woolsthorpe he had ground his own lenses, and it is possible that he prepared designs of his instrument then. We do know, though, that it was not until 1668, after his return to Cambridge, that he constructed his first reflecting telescope. It was a very small instrument, with a mirror about one inch in diameter, and six inches long; even so it showed all kinds of celestial objects such as Jupiter's satellites and the phases of Venus very clearly.

News of the telescope filtered through from Cambridge to London, and the Royal Society asked Newton not only for a description, but also for an instrument (it was their practice to test everything). Newton made a telescope for them, and in 1672 it was put to the test. The title of the description of the instrument is worth noting, because it shows the advantages one of the Fellows of the Society saw in it: "The new invention of Isaac Newton, Professor of Mathematics in the University of Cambridge; whereby long telescopes are considerably reduced in length without impairing the effectiveness of their use. Made public at the Royal Society, London 1671/72." [1] This underlies the fact that the days of the cumbersome refractor were numbered, and sets out yet another advantage of the reflector, an advantage that is still a very real one today.

Reports of Gregory's and Newton's designs spread through Western Europe, and around 1672 Jacques Cassegrain, professor of physics at Chartres in France, proposed yet a third design of reflector. Like Gregory, he used a main concave mirror with a hole in the center, but for his small mirror he had one with a convex (bulging) surface to reflect the light back down the tube (Figure 7–1c). Newton criticized Cassegrain's design for much the same

---

[1] The double date signifies that the document was prepared and the demonstration given sometimes between January 1 and March 24, 1672, because in England, at the time, the year officially began on March 25 and in Scotland, on January 1. Both dates were often given to avoid confusion.

reasons as he had Gregory's, though it did have a slightly larger field of view. It also had a very long focal length as the light passed along the tube three times and not just twice as in Newton's—this was an advantage because the tube could be shorter—but, like Gregory, Cassegrain did not build an instrument, and it was some years before his design was used.

Newton's reflector had many advantages and it was adopted by instrument-makers, who found the awkward long-tubed refractors becoming unpopular. But by one of those quirks of fate that seem to occur so often, the type of reflector most in demand and best made was based on Gregory's design, and this within a few years of Newton's death. One of the most famous of the many telescope-makers was the Scotsman James Short, whose Gregorian telescopes found their way everywhere from the 1730's onward. The instruments were not easy to build, for not only did Short and others have to grind and polish their lenses and mirrors, but they also had to mold their mirrors themselves, using a combination of different metals. This was the only way mirrors could be made, for there was then no known means of putting a film of silver or any other highly reflecting substance on the front of glass—this method was not invented until 1856. The metals involved were primarily copper and tin, though antimony and other substances were added, each maker having his own pet formula. In all cases the result was a highly reflecting metal that could be ground and polished, but that was very brittle and hence not at all easy to work with. Because it was used only for making a metal mirror, or speculum, the alloy became known as speculum metal.

To reach farther into space larger telescopes were needed, but in Newton's day, and for some time after, little concentrated work was done on the stars, so there was scarcely any demand for big mirrors. It was left to the amateur astronomer William Herschel to make the next move. By spring 1773 Herschel, a well-established musician in Bath in the west of England, had begun to make telescopes as a hobby. To start with Herschel built himself refractors, first of all making one about four feet in length; eventually he graduated to an instrument of thirty feet. Here he came up against the problem of mounting so long a telescope. Heavy metal engineering was only just beginning, owing to the introduction of cast iron and the development of the steam engine, so Herschel had to use the time-honored method of a pole, rope, and tackle. He found, as other astronomers before him, that the arrangements were inconvenient, and it was for this reason that he turned his attention to the reflector. He hired a Gregorian reflector two feet in length with an aperture of about four and one-half inches, and was so pleased with the way it handled that he began to see whether he could buy a large mirror, intending to fit it in a tube and mount it himself. To his surprise, he found that no one made large mirrors, so he could not do as he had for refractors and buy the optical parts, but had to grind and polish the metal disks himself. This was a formidable undertaking, but with the help of a book on practical optics and a collection of grinding and polishing tools, as well as some metal disks bought from another amateur astronomer, he set about producing mirrors. He placed orders for yet more metal disks and became so absorbed in the whole business of telescope-making that he spent every spare moment, including mealtimes, reading what to do and then, with the help of his sister Caroline and one of his brothers, Alexander, putting it into practice. He built dozens of reflectors, making more than one mirror for each and then selecting the best, and by 1778 he had completed a seven-foot-long reflector with which he went on to discover the planet Uranus three years later.

We have already seen that Herschel's greatest interest was observing the stars, and in 1781 he began to plan a really vast reflector. After considering various schemes, he decided to construct one with a length of thirty feet and a mirror of the then unprecedented diameter of thirty-six inches. No one could be found to cast such a large disk, so he decided to do the job himself in his own kitchen. After a series of experiments casting small mirrors with different combinations of tin and copper, in August 1781 a mold was made for the large mirror, and the metal heated and poured in. But the mold leaked a little, and the mirror was thicker on one edge than the other; to cap everything, it cracked when it cooled. A second casting was made, but this was even more disastrous; the mold broke its walls as the metal was poured in, and the molten tin and copper spread all over the kitchen floor, which was made of stone slabs. The heat cracked them and some patches exploded, sending stone splinters whizzing about the room; the workmen helping Herschel and his brother fled, but the two brothers remained, finally falling exhausted on a pile of rubble. Fortunately they were uninjured, but the accident stopped Herschel completing his scheme for some years.

By 1785 he had been granted his royal pension, and the optical quality of his small reflecting telescope was examined by other astronomers and declared as far better than anything they had come across before. The astronomer royal, Maskelyne, and others found that Herschel's claims, which had been criticized as exaggerated, were no more than sober truths. Now that he did not need to earn a living, Herschel was able to devote himself exclusively to astronomy and really concentrate on building a big telescope. He obtained a royal grant of £2,000, in addition to his pension, and this enabled him to begin. Though only $4,800 at today's rate of exchange, it was a very large sum of money for the time (a laborer only received about £25 a year, the astronomer royal lived well on £300, and Herschel's own annual pension was no more than £200). In addition the king gave him money for assistants— his sister Caroline whom he had trained was the chief one—and for workmen to help him. Before the project was finished, the royal purse provided a second £2,000. Large telescopes have always been expensive, and Herschel's was no exception.

The design had some novel points about it, not least the fact that the mirror was to be forty-eight inches in diameter and the tube forty feet long. No telescope even approaching this aperture had been contemplated before, let alone planned in every detail. Also, Herschel decided to do away with the second mirror Newton had used to bring the light to the side of the tube and, instead, tilt the main mirror so that the observer could peer down the edge of the tube—the diameter of forty-eight inches meant that the four or five inches blotted out by the head would not cut out a serious amount of light. The tube was mounted on a giant wooden framework with wheels (Figure 7–2), set on top of a brick wall sunk into the ground. This meant that it could be set reasonably quickly to face any point of the compass. At the bottom of the tube was a small hut, and here the observer's assistant sat, noting the results of observations (they communicated by a speaking tube). The

FIGURE 7–2. Herschel's giant reflector with a tube forty feet long and a mirror of forty-eight inches. The telescope was moved by winches and block and tackle, and the observer, who was at the front of the telescope, communicated with his assistant in the hut at the base of the telescope with a speaking tube. From E. Dunkin, *The Midnight Sky* (London, 1891). Courtesy of the Ronan Picture Library.

telescope was turned by winches, but was moved up and down by pulleys and ropes. It was an immense instrument and though Herschel had the two mirrors for it cast in London, the rest was his own work and design. It grew to be one of the sights that important people came to see when they visited the royal household at Windsor, for Herschel's observatory at Slough was only a few miles from Windsor Castle.

The giant forty-eight-inch reflector was completed in 1789 and Herschel used it to probe deep into space, though it was rather cumbersome, and for much of his work he preferred a twenty-foot-long telescope that he had built a little earlier. In 1839, seventeen years after his death, the forty-eight-inch was dismantled, yet it was not to be long before an even larger reflector was built. The new telescope-maker was William Parsons who, in 1841, succeeded his father and became third earl of Rosse. He was determined to take matters further than Herschel, and to build a larger instrument so that he could attempt to solve some outstanding problems that not even Herschel's monster could cope with. As Herschel before him, he gained practical experience by building small instruments, and in 1840 constructed a very successful reflector with an aperture of thirty-six inches, mounted in a similar way to Herschel's larger telescope. Rosse was encouraged by the instrument and began to design another with an aperture of six feet. Because of limitations in precision engineering, this seventy-two-inch giant was to be mounted between two brick walls so that, though it could swing freely in a north-south direction, its east-west movement was very restricted, and the observer had to wait for the earth to carry him round before he could see other parts of the heavens.

The telescope was built on his estate at Birr Castle, Parsonstown, in the county of Offaly, in what is now Eire. The casting of the first mirror began in 1842; after twelve hours heating, the metal was poured into its mold and this was taken to a giant oven where its cooling could be controlled over a period of sixteen weeks. But before grinding and polishing could start, the mirror was accidentally broken. Rosse set about altering the ratio of tin to copper so that it would be less brittle, and casting began again. The result was full of pits and holes and had to be discarded; on the fifth attempt the casting was successful. The large disk was subsequently cooled and safely transferred to the grinding and polishing machine that Rosse had designed and that was driven by a steam engine; the mirror was now in its last stages of production. All the stops and starts had taken time, but at last, in February 1845, it was completed. Rosse, Romney Robinson who had been acting as his assistant, and a well-known astronomer

James South were all ready to use it, having decided first to look at the hazy patch of light known as the nebula in Orion. They lined up the telescope and waited for Orion to be brought into position by the rotating earth, but just before the moment for observing, the sky clouded over; they had to wait for another twenty-four hours. When they did observe it the next night, the results were well worth waiting for, and much useful work was done with the telescope during the years that followed. This was just as well for the instrument cost Rosse £12,000 ($28,000), a fortune in 1845.

Yet, though Rosse's giant instrument, nicknamed the Leviathan of Parsonstown, was the most powerful tool for probing space ever constructed, its mounting did restrict its usefulness. "From the interruption by clouds, the slowness of finding with and managing a large instrument. . . . and the desire of looking well at an object when we had got it, we did not look at many objects," wrote George Airy, astronomer royal, and it became clear that new engineering techniques must be adopted to the full if big telescopes were to be really worth building. Beautiful and efficient mountings were already in use for much smaller reflectors, but the huge reflector presented a different problem. Two attempts were made during the 1850's and 1860's to improve matters. The engineer James Nasmyth, inventor of the steam hammer, built a twenty-inch reflector, with an elaborate mounting very like a naval gun support (Figure 7–3), an additional small mirror allowing Nasmyth to sit in the same position wherever he was observing.

The other attempt to use engineering methods was adopted by a brewer, William Lassell. He constructed a forty-eight-inch mirror and then asked Nasmyth to build a suitable mounting. The result was a hybrid between engineering techniques and the wooden contrivances of some time before because Lassell insisted that his reflector should follow Newton's design, which meant his observing position was near the top of the long tube, some thirty-six feet above the ground.

Nasmyth's and Lassell's telescopes represent the first important reflectors using the new style mountings and the last to have metal mirrors, for in 1856, the mathematician Karl von Steinheil and the physicist Léon Foucault began to make mirrors by depositing a very thin film of silver onto glass, using a method just

FIGURE 7–3. Nasmyth observing at his reflector, which was the first to have a fully engineered mount, and in which the observer sat in one position to observe anywhere in the sky. From S. Smiles, *James Nasmyth, Engineer* (London, 1883). Courtesy of the Ronan Picture Library.

invented by the chemist Justus von Liebig. Glass was easier to handle than speculum metal, and a few bubbles in it were of little consequence because it was only there to support the thin reflecting film on its surface and strong bars could be placed all the way underneath to prevent any sagging. Yet another advantage was that silver was a much better reflector than polished speculum metal.

One man to take up this new technique was George Calver, in England, who was commissioned by a rich businessman Andrew Common, to make an eighteen-inch silver-on-glass mirror for fitting to a tube and mounting designed by Common. Common used his new instrument with much enthusiasm, and in 1879 managed to take the first really successful photographs of Jupiter (Figure 7–4). He now decided to have a larger telescope, and Calver made him a thirty-six-inch mirror. With this he took more excellent photographs, and his success was owing to some extent to the way his telescopes were mounted. Herschel and Nasmyth had adopted what is now known as an altazimuth mounting, the telescope rotating in azimuth and moving up and down in altitude (Figure 7–5a). To follow any celestial object in its curved path across the sky the telescope has to make two movements. But Lassell and Common adopted the equatorial mounting, a style that Fraunhofer had used with his refractors. Many variations of the equatorial came into use, but its basic principle is shown in Figure 7–5b. The axis about which the altazimuth telescope pivots in azimuth is now tilted over, so that it is parallel with the earth's polar axis. In order to follow a star as it moves in a curved path, it is only necessary to move the telescope about its polar axis; the telescope tube will move in an arc. This single motion is ideal when one wants to take a long exposure as Common did, for the telescope can be driven by clockwork to follow stellar motions. This is why his photographs of Jupiter and of stars and nebulae were so clear, for he could give a long exposure, with the telescope automatically keeping in step.

Toward the end of his life, he set about building a reflector of sixty inches diameter, which was never successful; it was his thirty-six-inch reflector that was so notable. In 1885, however, he sold the sixty-inch reflector to Edward Crossley, an English north country

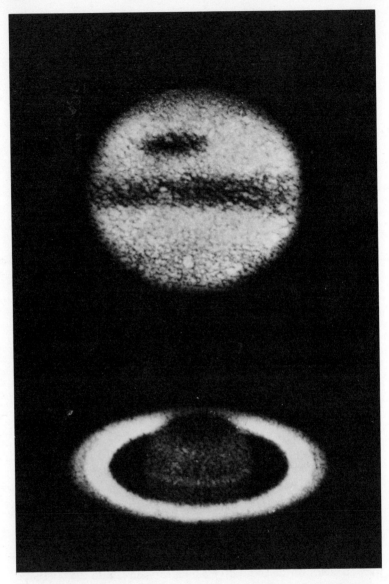

FIGURE 7–4. The earliest successful photographs of the planets
Jupiter and Saturn, taken by Andrew Common in 1879 and 1883,
respectively. From actual photographs inserted into the third edi-
tion of Agnes Clerke, *A Popular History of Astronomy in the Nineteenth Century* (London, 1893). Courtesy of the Ronan Picture
Library.

businessman who found the climate round him was too poor to make use of it, and the director of Lick in California persuaded him to donate it to the observatory. In 1895 it began work under the clear skies of western America and was, at the time, the largest reflector in the United States; with it James Keeler took some wonderful photographs.

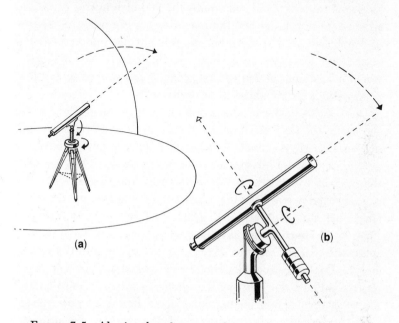

(a)   (b)

FIGURE 7–5. Altazimuth and equatorial mounts for telescopes.

Perhaps the most ardent big telescope astronomer has been the American George Ellery Hale, who was born at the time Lassell was using his forty-eight-inch metal mirror and who persuaded the financier Charles Yerkes to build the world's largest refractor with a forty-inch aperture. Hale was only twenty-five at the time, but no sooner had he succeeded with Yerkes than the new instrument persuaded him to seek out means for even larger telescopes to probe farther into space. Seven years after the Yerkes instrument was built, a new observatory was established on Mount Hamilton at the expense of Andrew Carnegie. Here, at the expense of another benefactor, James Hooker, Hale had a giant reflector built with a silver-on-glass mirror with a diameter of 100 inches. (See

Figure 7–6.) This had immense power and was able to photograph objects some 100 times dimmer than Herschel could pick out visually with his forty-foot instrument. Yet Hale did not even then rest content, and by 1928 he had conceived the idea of building a 200-inch aperture reflector and obtained a grant from the Rockefeller Foundation to finance it. In the event the telescope took a long time to design and construct—the 200-inch mirror presented all kinds of difficulties and the mounting set engineers all manner of problems—so that the telescope was not put into commission until 1949, ten years after Hale's death. It cost $6.5 million (it would cost double if built today), but it has proved and still is proving the far-sightedness of its designer.

One of the tasks of the astronomer is to make large-scale surveys of the sky, and here the giant reflector and refractor fail, not because they cannot do what is required, but because the area of sky they observe at any one time is so small that to cover the heavens would take a very long time. And time is short, because every big telescope is wanted for probing into the depths of space, not surveying areas. In 1885 Paul and Prosper Henry in Paris tried charting the sky by photography, and they were so successful that the director of the observatory, Admiral Mouchez, and the English astronomer David Gill worked out plans on an international basis for charting the whole sky photographically. The idea was that 20,000 photographs would be taken using special camera telescopes with a wider field of view than the ordinary telescope; the chart was to be known as the *Carte du Ciel*. It was a grand scheme, but some of the observatories who agreed to cooperate did not complete their works and the chart was never finished.

During the 1930's an invention by the Estonian optician Bernhard Schmidt, reignited interest in the reflecting telescope. Richard Schorr, director of Hamburg Observatory, encouraged Schmidt in his optical work, and in 1920 Schmidt moved to Germany. He was concerned about the limitations of the reflecting telescope, which only gave really good pictures of objects close to the center of its field of view. By 1929 Schmidt thought he had found the answer to the problem: a thin lens with a specially curved surface placed near the mirror so that light entering this new lens-mirror tele-

FIGURE 7-6. The James Hooker telescope at Mount Wilson Observatory, California, with a mirror with a 100-inch diameter. Courtesy of the Hale Observatories.

scope would give good images over a wide area of sky, typically some 5° by 5° compared with an area of less than 1° by 1° for an ordinary reflector. The Schmidt telescope, as it has come to be called, acts like a very fast camera lens, so it too picks up distant objects in space though it cannot do all that a big reflector can. Nevertheless, The National Geographic/Mount Palomar sky

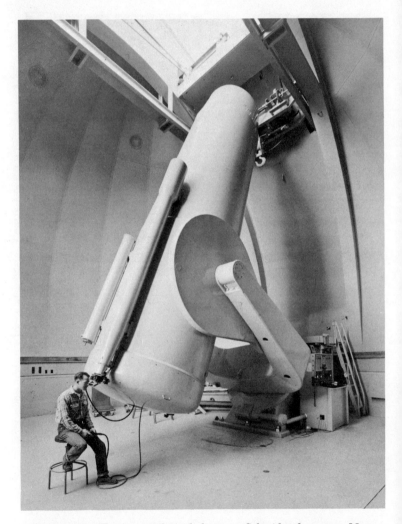

FIGURE 7–7. The forty-eight-inch diameter Schmidt telescope at Mount Palomar Observatory, California. Courtesy of the Hale Observatories.

survey made by the forty-eight-inch Schmidt telescope at Mount Palomar has been extremely useful to astronomers. (See Figure 7–7.) There is a twenty-inch instrument in Australia for surveying the southern skies, but appropriately the largest Schmidt currently in use is the fifty-inch at the Tautenberg Observatory in Germany. Schmidt died in 1935 but his optical method of using a thin lens—which is so much easier to make than a thick double lens of an ordinary refractor—has become important in astronomy, and instruments on this principle help the astronomer to make better use of expensive telescope time.

8 ✳ The Nature and Ages of the Stars

The stars are lights in the sky. This much was evident to the earliest observers, who seemed content to leave the matter there. Many thought the lights were carried by gods or spirits, or even that they were divine beings up in the sky, and only with the Greeks was their nature really discussed. Yet little progress could be made, because all that was known—or thought to be known— was that the stars were fixed and permanent, never undergoing any sort of change. In consequence, it was concluded that the stars could not be composed of the elements known on earth because all these underwent change and decay. The stars must be made of a divine substance—quintessence—and this put them beyond the scope of ordinary scientific inquiry.

Only after Western scientists had rejected the quintessence of the Greeks was it possible to reconsider the matter of the stars, but it was not until 1576 that a definite suggestion about the physical nature of the stars was made. Then Thomas Digges, in preparing a description of the new heliocentric universe of Copernicus for publication in a book called *Prognostication Everlasting*, showed the stars stretching out infinitely into space. He described them as "perpetuall shininge glorius lightes innumerable. Farr excellinge our sonne both in quantitye and qualitye." According to Digges, then, the stars were bigger and better suns, but what was the true nature of the sun? It had been worshipped as a god, and though the new philosophers from the sixteenth century onward would not countenance such an idea, they could say little more than that it seemed to be something burning away in space.

Ideas were still vague, even as late as the close of the eighteenth century, when William Herschel examined all the available evidence. The dark spots—sun spots—observed on the sun's disk were then explained in various ways. Some suggested they were scum, or solid bodies floating on the hot liquid surface, or the smoke from volcanoes, or even high ground uncovered by the ebb and flow of a bright fiery liquid sea. After considering his own observations, as well as those of astronomers since Galileo, Herschel concluded that the sun was a solid body, with a hot luminous atmosphere that probably contained a quantity of phosphorous, which helped to keep it bright. It "appears to be nothing else than a very eminent, large and lucid planet" and "is most probably also inhabited, like the rest of the planets, by beings whose organs are adapted to the peculiar circumstances of that large globe." Today his views seem wild, and hardly in keeping with his reputation as a great astronomer, but in 1794 they appeared a sane enough assessment of the evidence.

Is there phosphorous on the sun, and is this what makes it shine? The French philosopher Auguste Comte, who believed only in what could be found by direct, positive evidence, deemed this an unanswerable question. By 1844, when he was writing about astronomy, he stated quite definitely that, as man could not reach the stars, knowledge of their chemical composition was beyond his reach. This statement, though superficially logical, was really as brash as Hegel's pompous pronouncement that there could be no more than seven planets, and suffered the same fate. Even before Comte's view was expressed, work to solve the problem had begun. Indeed, during the middle of the eighteenth century, Thomas Melvill was investigating the way colors are refracted by a prism. Like Newton, Melvill was fascinated by light and colors, but he did not use sunlight because he wanted to measure each color separately, and therefore decided to use a series of colored flames as his sources of light. He generated the colored flames by burning such substances as sea salt in the flame of a methylated spirit lamp. He found a strange and unexpected result, for all his flames gave a distinctive yellow light that was always refracted by the same amount.

We can excuse Comte for not realizing the significance of Melvill's work; in fact almost half a century passed before anyone bothered to explore the subject more deeply. Then Melvill's experiments were repeated by Augustus de Morgan and William Wollas-

ton in England and Joseph Fraunhofer in Germany. All found
Melvill was correct. Wollaston and Fraunhofer decided that there
must be some difference in the flames if only they could discover
it. They realized that they must try to separate the colors as much
as possible, and because plain yellow flames did not seem to work,
they both hit on the idea of passing the light through a narrow slit
before it was allowed to reach the prism. The result was the same;
though both found that when they used sunlight instead of a
colored flame, the colored spectrum of the sun was crossed by

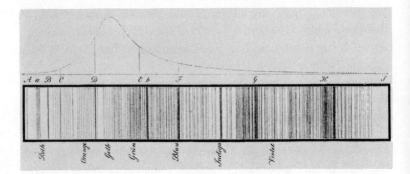

FIGURE 8-1. Fraunhofer's drawing of the sun's spectrum, with
red on the left and violet on the right. This shows the lines he observed and the letters he gave to the strongest ones. The faint
curved line above shows how brightness is distributed along the
spectrum; yellow is the brightest color because the sun is a yel-
lowish star. From H. Roscoe, *On Spectrum Analysis* (London, 1870).
Courtesy of the Ronan Picture Library.

dark lines. After this Wollaston lost interest, but Fraunhofer was
determined to retrieve something from what seemed to be a rather
fruitless set of experiments, and he concentrated his attention on
the dark lines. This led him to design and build a very efficient
spectroscope, and by 1815 he mapped the positions of 324 dark
lines. To identify some of the darkest and most noticeable of these
Fraunhofer lines he used letters, but as far as any explanation of
their cause went, he was still completely mystified. (See Figure
8-1.) Later he found that stars also gave colored spectra and that
these spectra were crossed by dark lines, but all he could finally
say about them was that they owed their origin to some factor he

could not understand, but that must have to do with the bodies themselves.

Fox Talbot, the photographer, working along the same lines, came to the conclusion that the presence of every color in the spectrum was owing to a chemical substance, but he could not prove this. He still found that when the metal sodium was heated and vaporized, it gave what Fraunhofer called the D line. The only trouble, as Fraunhofer had found, was that the D line was there when one heated other substances that did not seem to have any sodium in them. This was in 1826, and by the time Comte made his pronouncement almost twenty years later, no more progress had been made. Not until 1856 did a number of investigations suddenly begin to point in the same direction.

The Scottish chemist William Swan decided that the D line resulted from sodium and that it was present as an impurity in all the other substances tried by earlier experimenters; being a powerful source of light when vaporized it had been all that was noticed. Meanwhile, Léon Foucault found that if sunlight were passed through the bright light generated by vaporized sodium, the D line not only coincided with the yellow sodium line but was darkened. This led him to suggest that the vaporized sodium not only emitted the D line but also absorbed it when it came from another source. Another chemist, William Miller, uncovered a further vital clue. While working at King's College, London, Miller discovered that when he passed an electric current across two pieces of metal in order to generate an intense arc light, and used a spectroscope to examine the result, he obtained bright colored lines. And the important thing was that the colors of these bright lines were different for different metals. It seemed, then, that Fox Talbot might be right—every chemical substance emits its own particular colors when heated until it glows, the colors showing up as lines because of the slit that everyone used in front of the prism. What was now needed was an explanation of why this happened, and evidence to show why the sun and stars gave dark lines and vaporized materials bright ones.

Many chemists became fascinated by the problem, and Gustav Kirchhoff and Robert Bunsen, working at Heidelberg University in Germany in 1859, finally found the answer. By a host of brilliant experiments carried out with immense attention to detail, Bunsen and Kirchhoff established beyond all shadow of doubt that each chemical element emits its own characteristic series of lines and no others. The pattern of lines for each element is, in fact, its spectroscopic fingerprint. They also established two other laws that were vital for an understanding of the chemical and physical nature of celestial bodies. First, when a solid, a liquid, or a dense gas is heated until it glows, the spectrum it gives is continuous, going from violet to red. Second, when the light from a continuous spectrum passes through a thin gas that is cooler, the gas absorbs light just at those places in the spectrum that are characteristic of it. This was the explanation of the dark lines.

The spectra of celestial bodies have proved to be of vital importance to the astronomer. From them he can not only determine the chemical elements present in the outer atmospheres of stars, but, by studying the lines in detail, he can also find out many facts about their physical conditions, such as the temperature and pressure near their surfaces and the electrical conditions on them. Obviously the spectroscope could be a more powerful astronomical tool than anyone had realized, and it was only a very short time before astronomers were developing new techniques to obtain clearer stellar spectra. One of the most colorful characters among them, and the earliest in the field, was the English amateur astronomer William Huggins, who had the good fortune to be friendly with William Miller. Miller had great practical sense and realized the difficulties involved in obtaining star spectra good enough to show their lines in detail. Though he could easily generate his own light sources for analysis in the laboratory, to try to obtain a spectrum from a star even of the brightness of Vega, which appears 40 billion times dimmer than the sun, would be a monumental task. In fact he thought the chances of success were so slight that he tried to persuade Huggins to turn his attention to other problems in astronomy, but Huggins' mind was made up, and though his experiments proved long drawn-out, and tedious, he never lost heart. He began by making maps of the spectral lines of twenty-four chemical elements, and then hunted for the same patterns in stellar spectra, using his specially designed spectroscope at the eyepiece end of his telescope. As soon as wet-plate photography arrived Huggins applied it, even though it was so cumbersome. His observatory was rather like an obstacle course, because there was not only a telescope and a spectroscope, but also a large electric coil and batteries (so that he could photograph a laboratory spectrum against the star spectrum for comparison) and, to cap it all, a photographic tent for sensitizing plates before exposure and developing them directly afterward. Indeed, it was so cluttered that in some positions of the telescope, Huggins became jammed against the observatory walls!

But it was all worthwhile (see Figure 8–2), and in 1863, four years after tackling the problem, Huggins and Miller were able to present a paper at the Royal Society describing the spectral lines

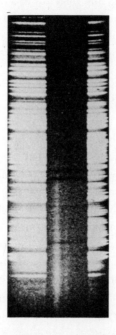

FIGURE 8–2. A photograph of the spectrum of a galaxy taken by Huggins. The galaxy spectrum is in the middle, with a comparison spectrum of iron above and below to allow the lines of the center spectrum to be identified. This photograph proved that a galaxy is composed of stars (see Chapter 9) and also shows that a star spectrum is composed of dark lines on a bright background as is the case, for instance, with the sun. Courtesy of the Ronan Picture Library.

to be found in some of the brighter stars, and identifying them with such chemical substances as sodium, iron, and hydrogen. In the United States, Lewis Rutherfurd had made similar observations but he seems to have become sidetracked by the possibilities of celestial photography. The next year, 1864, Huggins presented a more extensive paper to the Royal Society, but it was Fr. Angelo Secchi who tried to classify stellar spectra and, in 1867, had collected together enough observations to see some underlying system in them. Secchi thought there were four different types— (1) those that showed strong lines of hydrogen; (2) those with numerous lines owing to the vapors of metals; (3) those that showed bands rather than lines, with the bands sharp toward the red end of the spectrum and hazy toward the violet; and (4) stars with bands sharp toward the violet and hazy toward the red. Blue and white stars fell into the first category, yellow stars into the second, whereas the third was filled by orange and red stars, and the fourth by red stars only.

Secchi's classification was obviously very important as it clearly showed some differences between stars, though he was not able to explain what these might be. This was left to Norman Lockyer who, from his laboratory investigations (see Figure 8–3), correctly concluded that what Secchi had done was to classify stars according to their temperatures. He managed to do this because his laboratory work had led him to believe that all atoms, of whatever chemical element they might be, were conglomerations of the same particles, and in the classification Secchi had made Lockyer recognized just the kind of differences that different temperatures would cause if his atomic theory were correct. Lockyer also made spectroscopic observations of the sun, paying particular attention to the spectra of sunspots.

As sunspots appear like black dots on the sun's disk, some astronomers were convinced that they were holes in the sun's surface gases, which showed that the sun was cooler inside. Others thought that the coolness was owing to the downrush of gases from the surface, and in 1866 Lockyer found that the second explanation was the right one. He also discovered a new way of using the spectroscope. The famous French solar observer Pierre Janssen came across the same technique quite independently.

Lockyer and he were concerned with determining the nature of the pink flame-like prominences that could be observed round the sun's edge during a total eclipse, when the glare of the disk is cut out by the moon. For some time there had been an argument about whether they were something connected with the moon rather than the sun, but finally, in 1866, both men discovered these were hot glowing gases, mainly of calcium vapor or

FIGURE 8–3. Spectroscopic equipment used by Lockyer to investigate the sun's spectrum. The clockwork-driven mirror outside the window reflects sunlight through the lenses and into the spectroscope (left). The wires are shown connected to an apparatus with two iron rods to give an iron spark so that a comparison spectrum can be obtained. From J. N. Lockyer, *Studies in Spectrum Analysis* (London, 1904). Courtesy of the Ronan Picture Library.

hydrogen. The technique Lockyer and Janssen used was to widen the slit of the spectroscope and then allow only the edge of the sun to throw its light down the instrument. On some occasions they had to move the slit and telescope along the prominence, and their method was extended by Hale at Yerkes in 1889, where he developed the spectroheliograph. This was a spectroscope with two slits, one of which moved across the telescope's image of the sun's disk, the second one moving over a photographic plate so that it always isolated light of the same color. The result was that photographs could be taken of the sun in the light of a particular chemical element, and in this way a fuller picture of what was happening could be built up. Hale also studied the magnetic field of the sun, again using the spectroscope, for where there is a magnetic field the spectral lines are wider than usual.

Spectroscopic studies of the sun had made it possible to deduce what chemical substances it possessed and something of its physical conditions, but of course there was no independent proof that the deductions were right. There was still a little doubt in the minds of many astronomers about the whole business of interpretation. However, one of the facts that had come out of mapping the solar spectrum was that there was a series of lines that could not be identified with any chemical substance known on earth; the unknown element was named helium (from the Greek *helios* for sun), and for a time it remained a mystery. But in 1895, the chemist William Ramsay in London, isolated helium in the laboratory, and it was then generally agreed that the long awaited independent proof had arrived. Meanwhile those with faith in the power of astronomical spectroscopy as a means of investigating the nature of the stars pursued their work and, among other things, a more elaborate attempt was made to classify the many spectra available. In 1874 Hermann Vogel provided a new system that broke Secchi's method down further, but was this enough and what did it mean? Had it any use other than as a neat way of pigeonholing evidence? Most astronomers, or astrophysicists—for spectroscopic studies in astronomy became called astrophysics—thought that if they could discover a satisfactory classification of stellar spectra, they would have a deeper clue to the physical and chemical nature of the stars. This hope was what spurred on such men as Henry Draper at Harvard College Observatory, who had such enthusiasm that on his premature death in 1882, his widow endowed a research fund to enable his work to continue there. From 1886 onward, Edward C. Pickering led a team observing the spectra of all bright stars in the northern skies. One of the most prominent members of his staff was Miss Annie Cannon, and without her it is unlikely that they would have been able to publish in 1890 a new classification—the *Draper Catalog*. Here the different classes of stars were given letters instead of numbers, beginning with A and extending to Q, though Miss Cannon modified this eleven years later to include numbers, so that there would be ten subdivisions to each letter. As the years passed, this Harvard classification was amended, and between 1918 and 1924, a new *Henry Draper Catalog* was published. Here the old order A, B, C . . . was again changed and some letters dropped, so that astronomers have ended up with the following letter system: O, B, A, F, G, K, M, R, N, S—usually remembered by the mnemonic "O be a fine girl, kiss me right now, smack." Miss Cannon's numbers were retained so that, for example, the sun became a G0 star, Betelgeuse, M2, and Sirius, A1.

The Harvard classifications are certainly ones of color and temperature, and during the early twentieth century, even before the second classification had been derived, the results were used by the Danish astronomer Ejnar Hertzsprung and the American Henry Norris Russell. Hertzsprung was interested in measuring stellar distance and saw in the classification a clue to the true brightness of stars, because a blue star and a white star are hotter and therefore brighter over every bit of their surfaces than the cooler yellow and red stars. In 1905 he suggested a new idea, which he called absolute magnitude—the absolute magnitude of any star being the magnitude it would appear to have if situated at distance of 32.6 light-years[1] (that is, as if it had a parallax of 0″.1), and then proceeded to calculate the absolute magnitude for those stars whose distances were known. He next plotted the spectra of these stars against their absolute magnitudes, a method of displaying a relationship that was also adopted, quite independently, by Russell, and became known as the H-R, or Hertzsprung-Russell, diagram. Hertzsprung's purpose was to use the result to give him absolute magnitudes for stars of unknown distance by observing the spectra only, so that he could then calculate the true distances of the stars using their apparent magnitudes as well—what he called the method of spectroscopic parallax.

[1] To express astronomical distances in miles gives inconveniently large numbers that mean very little. An alternative is to use a larger unit than the mile, such as the light-year, which is the distance (not the time) light travels in one year. One light-year is equal to 6 billion miles.

Russell used the H-R diagram in a different way. He followed up suggestions made at the beginning of the century by Lockyer, who thought stars evolved from big cool bodies to smaller hotter ones, and then grew smaller and cooler again. Russell connected up this idea with the H-R diagram, incorporating both bright red stars and dim red stars that Hertzsprung had called giants and dwarfs as an indication of absolute magnitude, not size. His scheme was simple yet comprehensive. Stars began, Russell suggested, as large gaseous bodies that were fairly cool: they started, in fact, as N-type red giants. Under the gravitational pull of the mass of gas that composed the star, the whole body would contract and, as it contracted, so the star became hotter and moved across the spectroscopic scale from N through B. Russell thought that during this stage a star would keep to the same absolute magnitude because, though it became hotter and its surface brighter, so the surface shrank in size as the star contracted; the total amount of light remained the same. After the star had become a B0-type, it continued to shrink, now losing brightness. It moved down the sequence, this time from B0 through N, ending up as an S-type red dwarf.

There was much to commend Russell's theory, and when it appeared in 1913 it was generally accepted; but a decade later this was to change. By 1923 Arthur Eddington in England, who was deeply concerned with the internal constitution of the stars, was computing what conditions must be like below stellar surfaces if the ordinary laws of physics were to hold good. By considering the physical conditions of gases at the kind of pressure to be found inside a star, he discovered that the gases always behaved as if they were what the physicist called perfect: in other words, that they obeyed all the laboratory determined laws about pressure, temperature, and volume. But if this were so—and Eddington's figures seemed irrefutable—there was no reason why a star should begin to contract after it had reached the B0 stage; it should, in theory, continue on to become an O-type star. Was the H-R diagram really a diagram of stellar evolution at all?

The next step in finding an answer had already been taken by Albert Einstein in 1905, when he proposed his special theory of relativity. For the problem of stellar aging and development, the importance of Einstein's theory was its formula, which gave a relationship between energy and mass—the famous $E = mc^2$. The significance of this was that matter could be turned into energy, and the formula showed that if this happened, then the release of energy would be terrific—as we have now seen in the hydrogen bomb. Einstein's relationship provided a possible source of stellar energy for, by the turn of the century, it was clear that the earth was at least hundreds of millions of years old, and the question had arisen as to how the sun had derived enough energy to shine for as long as this. Energy from chemical reactions was nowhere near sufficient, and some other source was needed. Einstein's theory presented just the kind of possibility astronomers were looking for, but was it more than a theoretical pipe dream?

When in 1924 Eddington worked out the results that destroyed Russell's theory, it was still too early to say anything definite. Even though Eddington probably knew more about special relativity and the more advanced general theory of 1913 than anyone except Einstein, no definite method of energy generation could be worked out. This meant, too, that an alternative theory to Russell's could be no more than hinted at. But Eddington believed that matter into energy was the answer, and in 1932 his belief was supported by another English astronomer, Robert d'Escourt Atkinson, who was working at Greenwich Observatory. Using the new results of such physicists as Ernest Rutherford and John Cockcroft and Ernest Walton, who, in the same year, had together split the center core, or nucleus, of an atom in their laboratory, Atkinson worked out not only the energy released from possible atomic bombardments in stars, but even suggested the kinds of reactions necessary to produce the vast quantities of radiation required. This work was followed up, most notably by Hans Bethe, who, in 1938, computed other atomic reactions. With the additional experience of atom and hydrogen bombs behind them, and the immense amount of research into nuclear physics that has followed World War II, astronomers now have a greater insight both into the way stars evolve and into the time they take. Though there are a great many unsolved problems still facing them, their general opinion is that after stars condense from the dust and gas in space, they shrink from their original large size into red dwarfs, shining because of the nuclear reactions taking place deep inside them. Then they move upward in brightness and in spectral class. Gradually, the nuclear reactions build up helium atoms in their centers and the stars become unstable; they now begin to expand and brighten into giants. Finally, they shrink to become white dwarf stars and probably grow even dimmer. The time taken to complete their lives seems to vary and it is difficult to give exact figures: all the same there seems general agreement that the sun is probably about 6 billion years old, that some stars are older, some younger, and the total lifetimes must be measured in terms of billions of years.

9 ✳ The Nebulae

Stars and planets are easy to see in the sky, but only a very careful examination will show that there are also hazy patches of light. These are dotted about, and without a telescope less than a dozen can be seen in the northern skies, and not many more in the southern. Ptolemy cataloged five such objects, but the telescope has shown that three of them are collections of dim stars close together, though the other two do look like glowing clouds. When Ptolemy compiled his catalog, the few objects he mentioned were described as cloudy or misty, and subsequent catalogers noted the same kind of appearance, which is why they are called nebulae, from the Latin *nebula* meaning "cloud." What were they? When a sphere of fixed stars was the accepted picture of the heavens, the nebulae were thought to be holes in it through which heavenly light could be seen—heaven lying beyond the starry sphere. And even when the sphere of stars was cast aside, most astronomers still thought of heaven as lying beyond the material universe of stars, and they went on assuming that the nebulae were gaps in the most distant parts of the sky. Edmond Halley swept aside these ideas when in 1715 he wrote in the *Philosophical Transactions* that nebulae "are nothing else but the Light coming from an extraordinary great Space in the Ether i.e. covering great areas of space, through which a lucid *Medium* is diffused, that shines with its own proper Lustre." Put into twentieth-century language, Halley was proposing that nebulae were immense clouds of material that glowed on their own account. This explanation was so much ahead of its time that his contemporaries found it unacceptable; but time has proved Halley correct.

Though more nebulae were listed as the years passed, particularly by Charles Messier who published his large catalog in the *Connaissance des Temps* in Paris in 1784, it was not until William Herschel began to use his large telescopes, and particularly his forty-eight-inch mirror, that the nature of nebulae was investigated any further. In 1791 he had observed stars with hazy nebulosity around them, and there seemed to him to be two possible explanations: either the nebulosity was caused by dim stars appearing so close together that they could not be separated—in other words, that it was an optical illusion—or it might be caused by glowing matter, as Halley had supposed. A decade later Herschel had modified his views in favor of the explanation that most nebulae, including all the dim ones, were owing to clusters of stars, or stars like binaries and multiple stars in which three or four are orbiting about one another. Nevertheless, he had to admit that there was some milky nebulosity. As far as the nature of this nebulosity was concerned, Herschel was timid: "To attempt even a guess at what this light may be, would be presumptuous." All the same he did go so far as to suggest that it was possibly some kind of "phosphorical" condition.

But there was a new type of object connected with nebulosity that Herschel described, which he named planetary nebulae. These are not planets; Herschel adopted the name merely because in a telescope they appear as small greenish disks, just like the disk of the distant planet Uranus. He recognized these as stars surrounded by "dense luminous clouds," and argued against the idea that in this case they might be clusters of stars. However, the belief that some nebulosity might well be collections of stars remained in astronomers' minds. Herschel had proved many of the so-called nebulae to be of this kind when he observed them with his forty-eight-inch telescope, and there was bound to be some doubt about what an even larger instrument might make it possible to resolve into separate objects.

While these arguments based on observational evidence were going on, the theoreticians were busy speculating that the objects were clouds rather than collections of stars. In 1769 Pierre Laplace published his nebular hypothesis suggesting that the planets of the solar system had been formed from a nebulous disk of matter surrounding the sun. He believed that the nebula would shrink and rotate faster as it did so; he calculated that it would arrive at points when it was unstable and then throw off material that would contract into planets. In Herschel's forty-eight-inch telescope it seemed that there were a number of nebulae that, according to Laplace's hypothesis, could well be planetary systems in formation. The only way to decide seemed to be to use a telescope with greater resolving power—one of larger aperture. It was this that really lay behind Rosse's urge to build his seventy-two-inch telescope.

In early February 1845 Thomas Robinson and James South were able to begin using Rosse's telescope and they started by examining the nebulae listed by John Herschel in a catalog published in 1833. The results of these observations led them to sort the objects into three classes: (1) those that appeared round and were uniformly bright; (2) those that were round but showed more than one bright center or nucleus; (3) those that extended

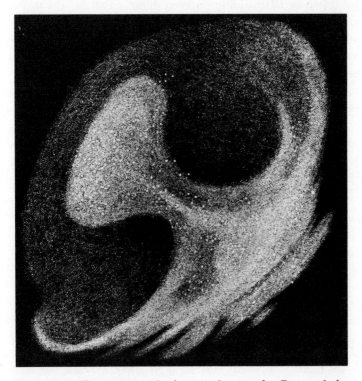

FIGURE 9–1. Engraving made from a drawing by Rosse of the Dumbbell nebula in the constellation Vulpecula (The Fox). He believed that this showed him that the nebula was made of separate stars. From A. Guillemin, *The Heavens* (London, 1867). Courtesy of the Ronan Picture Library.

in one direction, sometimes appearing as "long stripes or rays." They found they could easily resolve every object of the first class into stars; objects of the second class were more puzzling, but they displayed central stars with collections of "much smaller" stars surrounding them; the third class of nebulae seemed to be of the same kind as the second, but "being seen obliquely" were "therefore projected into ellipses sometimes almost linear." As a result, Robinson concluded that "no REAL nebula seemed to exist among so many of these objects chosen without bias: all appeared to be clusters of stars . . . ," and he wholeheartedly supported John Herschel's opinion that "a nebula, at least in the generality of cases, is nothing more than a cluster of discrete stars." In a paper published in 1848, Robinson even thought that he had managed to resolve most of the nebula in Orion into stars, and as this was a noticeable object and one that had previously given every appearance of being a true celestial cloud, evidence for the existence of nebulae seemed to be evaporating. Perhaps, after all, they were merely an illusion, owing to the inability of smaller telescopes to resolve the stars of which they were composed. (See Figure 9–1.)

The final solution of the mystery was due to William Huggins, whose pioneer work on the stars using the spectroscope was mentioned in the last chapter. The nature of the nebulae was a challenge he could not ignore, and in the second paper he and Miller published in 1864, he was able to report

> On the evening of 29th August, I directed the telescope for the first time to the planetary nebula in Draco. I looked into the spectroscope. No spectrum such as I expected! A single bright line only! At first I suspected some displacement of the prism, and that I was looking at a reflection of the illuminated slit. . . This thought was scarcely more than momentary; then the true interpretation flashed upon me. The riddle of the nebulas was solved. [See Figure 9–2.] The answer which had come to us in the light itself, read: Not an aggregation of stars, but a luminous gas.

So there was gas in space, and William Herschel had been right when he thought a planetary nebula was a star embedded in a vast shell of gas. But what of the other nebulae? In the next couple of years, Huggins examined the spectra of sixty, and of these he found that one-third were gaseous, giving bright lines, but that the remaining two-thirds had star-like spectra and must clearly, therefore, be collections of stars. What had been termed nebulae were really different kinds of objects—thin gaseous clouds extending, as Halley had said, over an "extraordinary great space," spherical shells of gas associated with a star and still called by Herschel's name of planetary nebulae, and collections of stars.

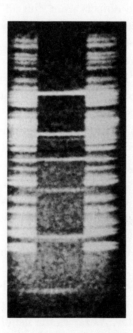

FIGURE 9–2. Spectrum of the Dumbbell nebula photographed by Huggins. As in Figure 8–2, the spectrum of the nebula is the middle one with comparison spectra above and below. It shows bright lines on a dark background, thus indicating a glowing gas, not separate stars. Courtesy of the Ronan Picture Library.

But the collections of stars were not simply a few stars gathered together; the objects seen in Rosse's telescope displayed an immense variety. Some were round, either clusters of stars or what, with even greater telescopic power, would doubtless be resolvable into clusters of stars. Then there were oval objects, and some that

appeared as strips or rays. Finally some of the nebulae appeared
to have bright centers and spiraling arms radiating out from them,
looking like giant pinwheels. All gave stellar spectra, and the
questions then arose of how large they might be and how far away
in space they lay. The direct method of measuring parallax proved
useless because it was confined to the nearer stars only, and these
star conglomeration nebulae were clearly farther away than that.
Hertzsprung's method of spectroscopic parallax (see Chapter 8)
could not help, except insofar as it confirmed that one must suppose that, because the objects were dim and required large telescopes to render them clearly visible, they must be extremely
distant. But the lack of precise figures did not, and never has, prevented careful speculation.

As early as 1750 ideas were proposed that were to have a
bearing on the problem. The man to put these forward was
Thomas Wright, a teacher of navigation and mathematics in
Durham in the north of England. Wright had his own telescope
and after some observing and much thinking, colored perhaps by
religious ideas, he suggested that the stars were not distributed at
random all over the sky. But if they were distributed in some
definite way, how could one tell this without knowing their distances precisely? Wright believed that the clue was to be found in
the Milky Way, which stretched across the sky in an irregular
band. He suggested this showed that there were more stars in
some directions than in others, the greatest density lying in the
direction of the Milky Way. As a result, he believed that the stars
were distributed in a kind of large disk in space and that the sun
did not lie at the center. As far as the nebulae were concerned—
and he classed them all together—he believed they lay outside the
disk; they were "external creations."

A somewhat similar view was arrived at in 1761 by Johannes
Lambert, who was unaware of Wright's suggestions. Lambert paid
particular attention to the irregularities in the Milky Way and
believed that it was moving in space. It was, he thought, only one
of many such systems; in this view the nebulae could be, as
Wright had termed them, external creations. It was William
Herschel, however, who made the most extensive investigations
into the stellar system, and in 1784 he published a paper in the

Philosophical Transactions, "Account of Some Observations Tending to Investigate the Construction of the Heavens." After concluding that the sun is not in the center of the Milky Way, which he calls a "great stratum" from an analogy with the geologist's layers of rock or strata, he then describes the Milky Way system, which he conceives of as an oblong box (Figure 9–3). The box, seen from inside as if projected on the inside of a giant sphere, will give the appearance of a band, with a split in one part; this, Herschel pointed out, was what the observer saw when he looked at the Milky Way.

Having discussed the great stratum of the Milky Way, or galaxy (from the Greek word for milk), Herschel considers the nebulae, which he felt at this time were stars not gas; he wrote "I soon found, that I generally detected them in certain directions rather than in others; that the spaces preceding them were generally quite deprived of their stars, so as often to afford many fields without a single star in it. . . ." From these and other details, Herschel concluded that there were other strata outside the galaxy and that the nebulae were evidence of them. Later on in his life, however, he felt less sure on this point.

When Rosse's telescope showed that some nebulae were spiral, and Huggins' spectroscopic observations made it clear that there were two kinds of nebula, the ground was already prepared for considering at least some to lie outside what we may term our own galaxy, though there were astronomers who had never accepted the idea of separate strata and still believed that all celestial objects were part of one large system. In fact, there was nothing definite to prove the existence of what amounted to separate islands of stars. It seemed to a number of astronomers that further investigations of the Milky Way might help in finding the answer. If our galaxy should prove to be a spiral like some of those observed by Rosse, and photographed successfully between 1898 and 1900 by James Keeler at Lick Observatory using Common's thirty-six-inch reflector, this might provide and important argument for the existence of separate galaxies.

Gradually observations accumulated, and studies of stellar motions made over a period of a dozen years by the Dutch astronomer Jacobus Kapteyn allowed him to announce in 1904 that

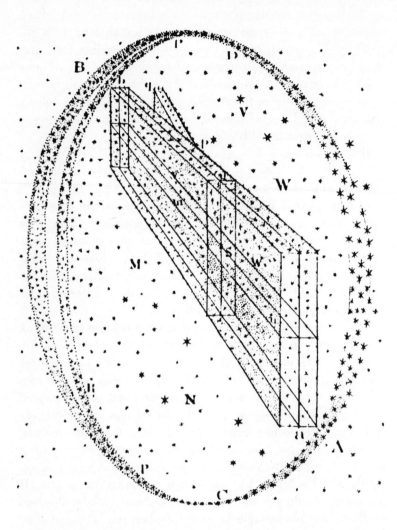

FIGURE 9–3. William Herschel's idea of the layout of the heavens. His belief that the stars were mostly collected together in a long rectangular shape was later found to be not quite correct and was changed to a disk of stars, which we now call our galaxy. From the *Encyclopaedia Londonensis* (London, c. 1807). Courtesy of the Ronan Picture Library.

there appeared to be two streams in stellar motions—one stream carrying some stars in one direction and another carrying most of the remainder in a direction that, when plotted on the celestial sphere, is almost at right angles to it. Considering the movements of the sun, Kapteyn's results could mean that there were two streams of stars moving in opposite directions. This was important for it might indicate that the galaxy was in rotation with one stream showing one edge of the rotating disk, and the opposite stream showing the other edge (Figure 9–4). Kapteyn developed a specialized method of statistical analysis to obtain his results, and his work stimulated others.

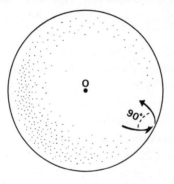

FIGURE 9–4. Motions of the stars in two streams as envisaged by Jacobus Kapteyn.

Finally, another Dutchman, Jan Oort, in 1928 put forward apparently conclusive evidence for the galaxy's rotation. Oort was able to give a figure of 250 million years for the time it took stars in the region of the sun to complete one revolution and to state that the sun did not lie in the center, which he believed to be toward the constellations of Scorpio and Sagittarius, in the densest part of the Milky Way. The fact that the galaxy was a giant rotating disk of stars and gas (the glowing gaseous nebulae) was established. A number of astronomers, most notably Arthur Eddington, were convinced that at least the spiral shaped nebulae were other separate systems similar to our galaxy. Space was full, they thought, of island universes.

But not all astronomers were willing to accept this view and, as late as the 1920's, there were many who opposed it. One argument they used was based on the distances of spirals, which were estimated from observations that were thought to show their rotation. The results led to the spirals appearing to lie well within the boundaries of our galaxy, the size of which had been calculated on the evidence of stellar magnitudes of very dim stars. Another argument was that the fact that such nebulae gave star-like spectra might merely mean they were additional bits and pieces of the Milky Way. How, then, was the matter to be decided? Obviously it was necessary to measure directly the distances of the spirals, and equally clearly there seemed to be no way in which this could be done. But in 1923 the matter was solved by the detection of variable stars of known luminosity in the nearer spiral nebulae (Chapter 11).

Once the distances of the nearer spirals were determined it was established, once and for all, that they lay outside the Milky Way system. There were found to be other Milky Way systems, other island universes, and for a long time they were called extragalactic nebulae, though now they are known simply as galaxies. Further observations of them were made photographically with the 100-inch telescope and with other reflectors, and the results carefully compared with photographs of our own galaxy. A number of interesting and important facts emerged (see Figures 9–5, 9–6, and 11–1). In the first place, it became clear that stars and bright gaseous nebulae were not the only occupants of spiral galaxies, or of our own. Photographs showed clearly that there were areas of dark gaseous material (dark nebulae) that were detectable because they blotted out the light of more distant stars. Later studies were to show that galactic space also contains a considerable number of dust particles; these also act as a blanket, blotting out the light from objects lying behind them. Similar dark patches are to be found in other galaxies, and there appeared to be every likelihood that our galaxy was a spiral too. Dust and gas were known to be the reason why we could not observe the central portions of our own star island.

But an examination of other galaxies soon made it clear that not all were spirals and that gaseous nebulae were not the only occupants of our galaxy. Some of the nebulae observed by Messier,

FIGURE 9–5. A spiral galaxy in Coma Berenices (Berenice's Hair).
This galaxy is viewed edge on, and dark obscuring matter is seen
stretching along its central regions (see also Figures 9–6 and 11–1).
Courtesy of the Hale Observatories.

Herschel, Rosse, and others turned out to be giant globular conglomerations of stars, or globular clusters. These are, in fact,
clusters of tens of thousands of stars, and they are components of
our galaxy. Studies by Harlow Shapley in the United States of
their distribution, as well as measures of their distances by
variable star technique, showed that they form a huge spherically
shaped system with its center close to the center of the galaxy.
Shapley's investigations, carried out in 1917, made it clear, too,
that the sun was certainly far away from the center of the galaxy,
though the figures he gave for the size of the galaxy have been altered over the years as more evidence about its dust and gas has
become available.

While Shapley was still doing his work at Mount Wilson, Edwin Hubble was probing the nature of the galaxies by making careful photographic examinations of the different types. He found that they could be grouped into three main classes: the spirals, which have already been mentioned, elliptical galaxies, and irregular galaxies. Elliptical galaxies (Figure 9–6) were so called

FIGURE 9–6. Elliptical galaxy, NGC 185. This is a small elliptical lying almost 2 million light years away and, like all ellipticals, seems to be composed almost entirely of stars with little or no dust and gas. Courtesy of the Lick Observatory.

because of their shape, for many of them looked rather like distant elongated globular clusters, though no separate stars could be seen. They gave spectra that showed them to be composed of stars; this, in fact, is how they were recognized, because many variations in shape from almost a sphere to very elongated ellipses were photographed. Irregular galaxies, like spirals, displayed dust and gas as well as stars, and they were so named because they ap-

peared to be lumps of extragalactic material with the character-
istics of neither spirals nor ellipses.

With the recognition of the existence of island universes at
great distances from our galaxy, man's concept of the universe
expanded. Admittedly he had for a time considered an infinite
universe of stars, but now he began to think of a universe composed of immense units, extending outwards to distances that still
stagger the imagination. It was a grander scheme of things than
he had ever thought of before, and, as we shall see now, it seemed
to be in continual motion.

10 ✳ The Red Shift

Motion in the stellar universe has always been difficult to detect. As we have seen (Chapter 4) it was not until the eighteenth century that evidence was given by Halley to show that the stars do not remain fixed in space, but have motions of their own. But Halley's discovery was concerned with one kind of motion only—motion across the sky—yet space extends in all directions, and there is nothing to make us suppose that stars may not move toward or away from the observer, as well as across his field of view. But motion to or from the observer along his line of sight—radial velocity—raises one crucial problem: how is it to be detected?

The difficulty arises because none of the customary ways used on earth for knowing whether a body is approaching or receding will work in space. On earth we see a body that is moving away from us seeming to get smaller as it recedes into the distance; conversely, a body that is approaching appears to become larger. In a telescope a planet's disk appears larger from night to night as it moves toward us during its orbit round the sun and appears smaller from night to night when it is receding, but with stars there is no detectable change. Stars are an exception to the rule because even the nearest are so extremely far away that they appear as no more than dots in the largest telescope; they display no disk and therefore there is nothing to get larger or smaller as motion moves them toward us or away from us along the line of sight. The radial velocity of stars is completely undetectable by ordinary telescopic observation.

In Vienna, in 1842, a principle that could help astronomers out of their dilemma was suggested by the German mathematician Christian Doppler. Doppler pointed out that if a source of light, such as a star, is moving toward an observer, the ups and downs of the light waves it emits will strike the eye more frequently than if the source is at rest. A simple experiment will illustrate the idea.

Suppose one has a piece of corrugated paper placed on a table top with its corrugations sticking up and that this represents a light wave in space. If the tip of a finger is placed on the piece of corrugated paper, and the paper moved slowly past the finger, one can feel the vibrations from the tops or crests of the corrugations. Now, if one repeats the experiment, but this time moves the finger-tip in the opposite direction to the moving sheet of corrugated paper, the vibrations will occur more frequently. The speed of motion of the corrugated paper has not changed, but the frequency of the arrival of the crests of the waves has. And if one experiences a higher frequency, this is just the same effect as one gets with a shorter wave, because in shorter waves the crests are closer together and are received more frequently than is the case with longer waves. So if a light source is approaching an observer he will receive a higher frequency than usual, and this will be interpreted as a shorter wavelength; in consequence the light will become bluer.

The opposite is also true. If a light source is moving away from the observer, the frequency of arrival of its waves will be reduced, and the crests will appear less often. This will give the effect of longer waves, and the light will appear redder. But by itself this Doppler effect is of little help; if a star is moving away and its light is redder, we shall not know this as there is nothing with which we can compare it. On the face of it we seem to be no better off for Doppler's discovery. However, in 1848, six years after Doppler published his principle, the French physicist Hippolyte Fizeau pointed out that not only would the color of the incoming light be altered by approach or recession of a star, but so would the position of its spectral lines. This was the clue needed.

When observations of stellar spectra are made, a laboratory comparison spectrum is observed at the same time so that the spectral lines can be identified. Ever since Huggins used this method (see Chapter 8) it has become standard practice. If, then, the spectral lines are shifted from their customary positions, this will show up because the lines of the comparison spectrum are normal. There are all kinds of refinements necessary; the photographic plate is exposed to the comparison spectrum before it is exposed to the star, and then again afterward, so that any changes in the spectrograph owing to cool air, or any other cause, are shown up by a slight change in the position of the comparison lines. Any such change can then be taken into account when changes in the positions of the star's spectral lines are measured. Using the Doppler principle as interpreted by Fizeau—what should, in all fairness, be called the Doppler-Fizeau effect—a body that is emitting light and moving toward us should present a shift of its spectral lines toward the blue end of the spectrum. And if it is receding from us it should present a displacement of the lines toward the red end—a red shift. What is more, the amount by which the lines are displaced is a measure of the body's velocity.

So far the matter of blue and red shifts of spectral lines have been considered theoretically. But do they exist in practice? Have they been observed and, equally important, been observed in the case of an object that we know is moving? Binary stars orbiting about each other present an obvious testing ground, and shifts of the kind predicted by the Doppler effect have been observed. This in turn has led to the detection of binary systems where the stars are too close to be resolved as separate points of light, or where the binary system is too far away to permit even a wide double to be observed. Such systems are called spectroscopic binaries, and Huggins made the first measurements of their radial motion.

The sun is another obvious case where a Doppler shift ought to be observed because it is rotating, and presents the astronomer with a big enough disk for him to compare the lines seen at one side and those seen at the other. Because of rotation, one side will be approaching the observer and will display a blue shift, and the other side will be receding and will give a red shift. In June 1871 Vogel managed to obtain spectra with shifts of this kind, shifts that, as one contemporary put it, constituted "at once the test and the triumph of the method." And if further evidence were needed, blue shifts were also detected in the huge flame-like prominences as their glowing gaseous material was thrown upward from the sun.

For the study of stellar motions and, in particular, for analyzing the motion of our galaxy, knowledge of radial motions is vital. Jan Oort made much use of this knowledge when calculating how

stellar motions could be fitted into a pattern and concluding that
the galaxy is in rotation. But the most spectacular effect of radial
motions has been in the problem of deciding about the nature and
behavior of other galaxies. To begin with, no radial velocities
were observed. In 1874 Huggins examined six nebulae but could
detect no shift, blue or red, in any of them. Between 1890 and
1891 Keeler tried again with a telescope larger than Huggins';
this time he was extremely successful. He found shifts for ten
planetary nebulae (one in Draco showed a radial velocity of
forty miles per second toward the earth) and that the great
gaseous nebula in Orion was receding at a velocity of some eleven
miles per second, though it was realized that this might only be an
apparent motion resulting from the motion of the sun and the
whole solar system with it. Radial velocities for most of the galaxies were more difficult to come by, and it was only after 1912
that definite evidence was at last found. Between 1916 and 1917
Vesto Slipher at Lowell Observatory, Flagstaff, Arizona, and F.
Pease at Mount Wilson, managed to obtain some forty galactic
radial velocities, and these varied from 500 to 1,100 miles per
second; the strange thing about them was that almost all showed
red shifts: there was hardly a blue shift among them. (See Figure
10–1.) As Harlow Shapley remarked in 1919, there appeared to
be no straightforward explanation of this peculiar fact. But was it
real? Would blue shifts show up when even more galaxies had
been examined?

When the 100-inch reflector came into use in 1919, one of its
tasks was to determine radial velocities for galaxies whose spectra
were too dim to be observed in smaller telescopes. More spectra
were photographed, especially by Milton Humason, and even
greater velocities of recession were found, the fastest being a galaxy in the constellation of Gemini that was moving away at a
speed of 15,000 miles per second. But again there were no blue
shifts; it seemed as if the whole universe of galaxies was expanding, just as particles of gas would if released from a container and
allowed to disperse into space. However, the velocities of recession
using the Doppler-Fizeau interpretation of red shifts were nothing
less than phenomenal. Even those astronomers most strongly committed to such an interpretation were staggered at the fantastically

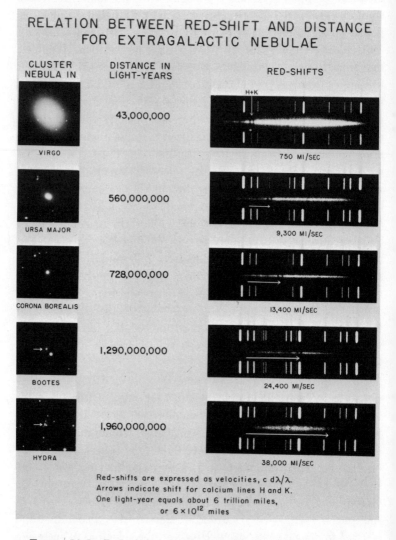

FIGURE 10–1. Red shifts of galaxies, showing how the shift increases the farther away a galaxy is. This is the reason that the more distant galaxies are said to be moving more quickly than those nearer to us. Courtesy of the Hale Observatories.

high values, which far exceed the movements of even the most energetic stars. Could they really be right? And if the amounts of shift observed were correct and there seemed no possible doubt about them, could they result from some other cause? Was recession perhaps the wrong way to interpret them?

The theory of relativity seemed to provide some possible alternatives, but before detailing them we must spend a moment on the theory itself. Einstein's work arose from theoretical investigations and was supported by experiments conducted in 1881 and later by Albert Michelson and Edward Morley, which were the result of arguments about how the earth's motion affected light waves received from objects in space. This was a natural problem for physicists to consider, because it had recently become known that light not only gave optical effects through lenses, but also that its rays could be affected by magnets and by electricity; light was, in fact, an electromagnetic radiation. Light waves must be investigated in every way, and with the knowledge of the Doppler effect it was only to be expected that the motion of the earth should come into the problem. And there was a particular reason for this—the idea of light waves and of their electromagnetic effects made it seem clear that the waves must travel in something. It just did not seem possible to have a wave all on its own; it must be a disturbance in some substance—it must have something to wave. The something could not be air, because light waves were observed to travel just as readily through a vacuum; physicists hypothesized the existence of a transparent weightless something, which they called ether.

The next project was to find out more about this ether and, in particular, how it moved with reference to the earth. To determine this Michelson set up a delicate experiment, using an optical arrangement of mirrors and a powerful light source. He split the light from the source into two beams, one that was allowed to go out and be reflected back in the direction in which the earth was moving and one that was turned through a right angle and then allowed to go out and be reflected back. He then measured the time taken by the two beams to go out and be reflected back over precisely the same distances. Michelson expected to find a difference because in the direction of the earth's orbit the beam would travel at a speed affected directly by the speed of the earth's motion, and at right angles it would travel at a speed affected only by a fraction of the speed of the earth's motion. Michelson first performed the experiment by himself, but he could detect no difference. Admittedly, he expected the difference to be very small and hard to measure, because the speed of light is 10,000 times the speed of the earth in its orbit, but he believed his apparatus was delicate enough to allow him to make the measurement. He therefore set about getting assistance; with Morley he devised a new and extraordinarily delicate experiment at the Case School of Applied Science in Cleveland in 1887. But again no difference could be detected. Most physicists felt that there was something wrong with the experiment and urged Morley to try again. Morley made a further attempt in 1904, but still he could obtain no measurable result.

The experimental evidence certainly seemed to be nonsensical, for it appeared that if one moved relative to a beam of light, this gave no difference to its measured velocity compared with if one stood still. Various explanations were offered, the most acceptable being that put forward by the Irish physicist George FitzGerald in 1892, after the Michelson-Morley joint experiment. FitzGerald suggested that all moving bodies shrink a little in the direction in which they are moving. For instance, an arrow flying through the air would shrink a small amount so that it would be shorter than when stationary, its breadth and thickness remaining the same. In the Michelson experiments the contraction would affect the scientific apparatus used, and so make nonsense of the measurements. But ingenious though FitzGerald's explanation was, other questions forced physicists to reconsider the whole problem of motion, not only of light waves, but of everything in the universe. There seemed to be no place to which motion could be referred; one could think of the earth as stationary if one liked, or one could think of the earth as moving and the sun at rest, but if the galaxy moved also—and it was unlikely to be the only one in the entire universe to be still—then where should one find a stationary place? Newton's laws of motion, which state that a body is either in a state of motion or a state of rest, demand some fixed point.

The French mathematician Henri Poincaré, in 1899, came to

the conclusion that there was no fixed point in the whole universe, and so all motion must be motion relative to something else. One cannot speak of something having an absolute undeniable motion, or being utterly and completely at rest. If a body moves at a given speed relative to the earth, at the same time it is moving at a different speed relative to the sun. And if a body is stationary with respect to the earth, it is certainly not stationary with respect to the sun, to any other star, to our galaxy, or to any other galaxy. Motion and rest were only relative terms. But as Poincaré pointed out, all laws of science had been determined relative to the earth, considering the earth as stationary. New laws were, he believed, now required, laws that would stay the same whether a body was in motion or not relative to any other body.

Poincaré's principle of relativity explained the Michelson-Morley experimental results, for if every body in space is always in relative motion, FitzGerald's contraction will always be occurring and no experiment can ever show the difference Michelson had originally expected to find. In other words, the velocity of light will always appear the same, whether the observer and his apparatus are standing still or not. And this still holds, even if we do away with the ether, as we do today, for light is now thought of as packets or quanta, of waves. This new view does not affect the Doppler-Fizeau principle, for we are there concerned with counting the frequency of wave crests and obtaining the effect of a different wavelength; the velocity of light is taken to remain the same all the time for, if it did not, then the red shift would not be seen.

The reinvestigation of the laws of physics in the light of new evidence provided by Michelson's experiments was carried out by a young German-born Swiss, Albert Einstein. He took as his basic principle the constancy of the velocity of light, however an observer may be moving relative to any source from which the light is coming. On this he based the rest of his deep and penetrating work, developing a rigid mathematical system that made the laws of physics the same for all observers on bodies moving relative to one another with a fixed velocity. He published his results in 1905. A storm of argument and assessment arose, not all physicists accepting the fundamental changes that the mathematical results demanded. Meanwhile, Einstein worked on, extending his special theory, which dealt with observers in uniform motion with respect to one another, to a general theory in which observers with changing velocities could be accommodated. The mathematics was more difficult, but the theory he announced in 1915 brought us closer to the universe as we observe it to be.

One of the upshots of the theory of relativity was the relationship between energy and mass (see Chapter 8), which was to prove the clue to the way stars generate their power. Another was the fact that the orbital motion of the planets proved to be slightly different from that calculated on the basis of Newton's theory. The general result of the two methods was the same, but in matters of fine detail differences did show up, and this was particularly so in the case of Mercury. Mercury has the most eccentric orbit of all the main planets excepting Pluto, and it had long been known that this oval orbit, like those of the other planets, rotated around the sun. The amount of rotation each century is negligible for most planets, but in the case of Mercury it amounts to 9' 34". Of this figure, the greatest amount of the rotation results from the gravitational pull of the other planets and, as computed on the basis of Newton's gravitation theory, amounted to 8' 52". Thus a motion of 42" per century was unaccounted for by Newtonian principles, a discrepancy that had worried many astronomers, especially Leverrier. But if planetary motions were computed by Einstein's theory instead of Newton's, there should be an extra factor causing a motion of an additional advance of 42" per century in the orbit. An observational test gave a result of 43", which was so close to the 42" discrepancy that, considering the probability of a small error in the observed measurements, it was taken as a proof of the correctness of general relativity.

Another very significant proof concerned the nature of light. According to Newton's ideas, a ray of light will be slightly bent from its straight path when it passes close to a massive body. But on relativity theory this bending will be greater than if computed on Newtonian theory. The total solar eclipse of 1919 presented the first opportunity to test Einstein's theory on this point, because during such an eclipse the stars that appear close to the sun in the sky will have their light beams distorted by the sun's massive gravitational pull. It was therefore important to determine how large this bending might be, and Arthur Eddington went to Principe in the Gulf of Guinea off West Africa to make measurements, while other astronomers visited different observing sites over which the moon's shadow was to pass. On Newtonian theory the bending should amount to 0".87, and on Einstein's relativity theory to 1".75: Eddington's observations and those of his colleagues at an observing station in Brazil gave figures that were very close to Einstein's and very far from Newton's.

Lastly, there was another consequence of relativity theory, which directly affects red shift measurements. The movements within an atom give rise to its emitting light at a certain wavelength but, according to relativity, this wavelength will vary if the gravitational conditions vary. In other words, an atom of sodium in a laboratory on earth will emit the D lines, and an atom of sodium on the sun will also emit D lines, but those from the sun will have a slightly different wavelength from those on earth because the gravitational pull of the sun is much stronger. The effect is noticeable as a red shift. The amount of shift in the solar spectrum is very small, but red shifts had been noticed as early as 1897 by L. Jewel, and in 1909 by Charles Fabry, but they had been explained as results of pressure in the surface gases, and of motions of the gas, giving a Doppler effect. Was this the cause, even though the shift was about that to be expected from relativity? Observations made in 1923 and 1924 still did not seem conclusive, and an independent observation of a different kind was really required. Yet it was not until 1926 that Walter Adams, using the 100-inch telescope at Mount Wilson, observed the shift in the spectrum of a white dwarf. The star he used was the binary companion to Sirius and was remarkable for its immense density, which was estimated to be of the order of one ton per cubic inch. It was an excellent choice of object and acted as yet another confirmation of relativity.

In the light of Adams' observations and of relativity, did the Doppler interpretation of the observed red shift of the galaxies stand? Could they be interpreted as a real recession? Calculation showed that if one made some assumption about size, and then worked out the mass of the stars involved, the red shift should not be the immense value that was observed. For white dwarfs the shift was a useful cross-check that the star was a dwarf, but it could not be used to replace the Doppler effect explanation. However, one of the consequences of general relativity, worked out in 1917 by the Dutch astronomer Willem de Sitter, was that the most remote objects in the universe ought to move away into space. This was a kind of force, the opposite of gravitation, which only took over when the universe was empty—which we know it is not—but the discovery of the red shift of the galaxies made it necessary to pursue the matter further. Another effect De Sitter obtained from relativity was that light of any given wavelength would appear to be of a longer wavelength to an observer situated at an immense distance from the source. For such objects as galaxies, which are so very far off, we ought then to have a red shift resulting from this optical effect. But later studies of relativity made it clear that the optical red shift was really all part of the recession of distant objects—cosmical repulsion—and so it seemed that the observations of receding galaxies fitted in well with relativity theory.

By the 1930's the idea that all galaxies were moving away from one another was generally accepted, and what Eddington called the "expanding universe" came to be part of astronomical thinking. Clearly the whole universe—or at least as much of it as can be observed—was undergoing some kind of change; galaxies were constantly moving apart, and the space between them becoming emptier. This was the picture in 1933, but since then many new observations have been made, and new ideas and interpretations have been posited. Yet the reality of the outward motion of the galaxies is not in question now, and the red shift truly seems to be a Doppler effect.

11 ✳ The Size of
the Universe

The size of the universe is another question that has intrigued man from the moment he began to look at the sky. Again there have been various answers depending on the extent of the astronomical knowledge of the time. For long, astronomy was inextricably interwoven with man's religious ideas and his beliefs about how the heavens affected life. The general trend has been to realize that it is larger than was previously believed; we must now look into this a little more closely, for it seems that even our current ideas are undergoing rather startling changes.

When the universe was no more than a dome of stars with a flat earth beneath, most men had not traveled very far from home. The few that had gone a long way had migrated to distant lands, and all they knew was that they had voyaged for a long time. Their world, the earth on which they lived, was comparatively small, and at best they could think of the heavens as lying as far as the most distant journey. Even when they had developed a measure equivalent to the mile, their systems of counting did not contain the vast numbers we have today, because they did not have any use for them. Even the measures made by Aristarchos did not give all that large a distance for the sun and moon, even though the size of the earth was known reasonably well. It is difficult, if not impossible, to express in miles how big the Greeks thought the universe to be, but by the time of Hipparchos they believed the sun to lie at a distance not greater than some 4 million miles, so we can conjecture that they considered that the stars were not farther than a few million miles. To be more precise would be to give figures where no figures existed—they realized that the stars were much farther away than the sun, and the figures for the sun were immense, especially when we realize that Greece itself covered no more than 300 miles from north to south.

Copernicus' universe was essentially different from the Greek universe in size. There was a sphere of stars, but the sun was now at the center, though (see Chapter 2) no shift of the stars was observèd thoughout the year and, if the earth did not move in an orbit round the sun, this meant that the stars must be immensely farther away than the Greeks had ever imagined. Tycho Brahe could measure correct to 1' and so they must be farther than 300 billion miles or, in modern terms, farther than one-third light-year. To some this seemed nonsense, but the idea of such immensity caught on among astronomers, especially after Thomas Digges did away with the sphere of the stars and replaced it with an infinite universe. Halley discussed the matter a century and a half later and as a first step assumed that the stars were evenly spaced throughout the universe; he then worked out how many stars there must be at various distances, each increase in distance being computed on the assumption that all stars were roughly the same brightness, so that magnitude was an indication of distance in space.

William Herschel was intrigued throughout his life with the problem of determining stellar distances (see Chapter 6) and his attempts to measure parallax led him to the discovery of binary stars. His failure to detect stellar parallax did not stop him attempting to find the extent of the Milky Way system, and in a paper read before the Royal Society in June 1817, five years before his death, he summed up his ideas on "The Local Arrangement of the Celestial Bodies in Space." Taking as his basis Halley's ideas of the even distribution of the stars—only space in this respect meant the Milky Way—and his and Halley's approximate result that brightness is a measure of distance, he set to work calculating the distances of the dimmer stars he could observe with his forty-eight-inch telescope. His assumption that all stars are of equal brightness was, of course, attacked, but, as he remarked, it was no different from taking an average height for a number of men chosen at random—probably none would be of average height but, by and large, this represented how tall they were. Over the question of the even distribution of stars, the

existence of double stars and, more especially of the clusters of
stars he had detected with his telescope, made it clear to him that
stellar distrubtion was not really even but a matter of averages.
On the average, making such an assumption did not violate observation too badly.

Herschel's investigations led him to conceive of the Milky Way
as extending more than 2,300 times the distance of a bright star
such as Sirius. Sirius, he knew, had a parallax of less than 1″, so
the extent of the Milky Way system must be at least 45 million
billion miles or, as Herschel put it, fathomless. Yet this figure of
7,500 light-years is far too small according to our modern
measurements, and the extent of our galaxy is very much greater
than Herschel imagined. But what of the nebulae, and especially
the extragalactic nebulae, if they existed? Herschel gave no
figures, nor could Rosse, and it was not until 1885 that a new
possibility presented itself. In that year a supernova appeared
near the center of galaxy M 31 (number 31 in Messier's catalog),
the great spiral in Andromeda. (See Figure 11–1.) The star
attained so great a brightness that it gave an apparent magnitude
of seven, that is, it accounted for more than one tenth of the light
from the whole galaxy. But although some suggestions were made
for determining the distance of the galaxy from the sudden flare-
up of such a star, during the 1880's knowledge about just how
bright a supernova could be was very limited; not until 1917
could the idea be pursued further. Two novae, less extraordinary
but equally important, were then observed in the same galaxy,
but, more significant, George Willis Ritchey at Harvard Observatory found a nova in the spiral galaxy NGC 6946 (number 6946
in the *New General Catalogue*) and with a colleague began to
hunt through past photographs of galaxies to find evidence of
other novae. A discussion then arose with Harlow Shapley, who
concluded that if the novae were actually as bright as those observed in our own galaxy, then the Andromeda and other galaxies
must be near enough to be considered parts of the Milky Way,
having distances of some 20,000 light-years. If they were to be
considered as spirals on their own account, comparable in size
with our galaxy, then their distances would be such as to require
the novae to be far too intense. But Shapley's argument was based

FIGURE 11–1. The great spiral galaxy in Andromeda, M 31. Lying about 2.2 million light-years away, it is comparable in size and shape to our own galaxy. Courtesy of the Hale Observatories.

on measures of the absolute magnitude of novae made by van Maanen in 1929; it later became clear that there were, in fact, two types of nova, the ordinary nova of which perhaps between twenty and thirty appear every year in our galaxy alone, and the rarer supernova, which may attain a brightness some 5,000 times greater than a nova, but of which there is only likely to be one every few hundred years. With the possibility of the existence of supernovae, and it seemed that this is what the 1885 object in M 31 was, Shapley's argument lost much of its force. All the same, it had not been proved wrong, and what was required was some new way of determining distance.

Henrietta Leavitt found the long-sought clue in 1912 when working at Harvard Observatory. During the previous four years she had studied a host of photographic plates taken in Arequipa, Peru, where a Harvard telescope had been set up; she paid particular attention to photographs of the two Magellanic Clouds. Here she found a number of variable stars that changed their light over a matter of a few days or a few weeks, but for the most part having periods around five days. This in itself was not so surprising, but these stars behaved in an unusual way, for their average brightness—between maximum and minimum—depended on their periods. For those with a period of two days, the mean magnitude as observed in the clouds was 15.5; for those stars having a five-day period the mean was 14.8; for a ten-day period, 14.1; and for a 100-day period, 12.0. This period-luminosity relationship was to prove the vital link in the chain of distance determination.

Such a period of variation was typical of the five and a half day variable star δ (delta) Cephei, and these short-period variables have become known as Cepheids. Their discovery caused a considerable stir, because they could obviously be looked on as a kind of lamp of standard brightness, so that if one could determine the distance of a nearby Cepheid in our own galaxy, then it would be possible to find its true luminosity or absolute magnitude, and from this determine the absolute luminosity of the Cepheids in the Magellanic Clouds. Working out the loss of light with distance, the distances of the Magellanic Clouds could be found. What is more, if Cepheids could be discovered in other galaxies, then here would be a certain way of determining distance.

Hertzsprung, in 1913, was the first to fix the scale of absolute magnitudes by observing the motions of thirteen Cepheids and then working out their mean parallax. There were some assumptions to be made, but at least it was a start; he gave the distance of the smaller Magellanic Cloud as 32,000 light-years. However, this figure was not accepted for long, as it became clear that the apparent magnitudes of the dimmer stars in the southern skies were wrong and, after they had been revised, so were Leavitt's assessments of the magnitudes of the Cepheids in the clouds. The distance of the smaller Magellanic Cloud was then recomputed by Shapley, who in 1918 announced that its distance was 94,000 light-years. This made it clear that it could no longer be considered a piece of the Milky Way; it lay beyond our galaxy. So did the larger Magellanic Cloud at 84,000 light-years.

Shapley made great use of the Cepheids, applying them to the globular clusters, many of which he had discovered in photographs he had taken between 1916 and 1917. He found the distance of the nearest, which is in Centaurus, to be 21,000 light-years, and the most remote 220,000 light-years, though the latter, as well as other more distant ones, were estimates, because Cepheids could not be detected and Shapley had to assume that all globular clusters were similar in size and brightness. Now he was ready to tackle the question of the spiral galaxies.

Shapley was clear in his own mind that they were parts of our galaxy, not separate objects, and he cited in support of his contention the fact that all appeared clustering fairly close to the poles of the galaxy. They did not lie all over the sky, as one would expect if they were external islands of stars, but above and below the main flattened system of stars and nebulae that composed our galaxy. But gradually, as better photographs were taken by the large American reflectors, especially those at Mount Wilson, and sky survey photographs were multiplied at Harvard Observatory, two facts emerged: the galaxies displayed dark material in their central and their flatter regions, and more galaxies were found, many of them lying close to the Milky Way. Thus it became clear that the clustering of galaxies near the north and south poles of our galaxy was not a real effect; it was a kind of optical illusion, owing to the fact that not only the stars of the Milky Way, but also the associated dark nebulae and dust blotted out many galaxies. They were there but could not be seen.

Finally, in 1922, J. Duncan announced that he had detected a variable star in a spiral galaxy and, within a year, Edwin Hubble and colleagues at Mount Wilson had found a Cepheid variable in the Andromeda galaxy. Then Cepheids were found in other nearby galaxies, the spiral M 33 and the irregular galaxy NGC 6822, and it became evident that their distances should be measured in millions not thousands of light-years. The extragalatic nature of the galaxies was no longer in doubt.

In the last chapter we saw that the application of the Doppler effect led to the discovery that all galaxies displayed red shifts. The observational side of this was examined and extended by Hubble, who in 1929 found that whenever he compared the red shift and the distances determined either by the presence of novae or by Cepheid variables—and both methods had been proven to give similar results—they were related. On the face of it the relationship was a strange one, for the farther away a galaxy lay the greater its red shift. In other words, the more distant galaxies were receding more quickly than those nearby. By 1933 enough distances and red shifts had been determined to give a figure for the relationship between distance and shift, and it was widely accepted to be between 500 and 1,000 kilometers per second per megaparsec.[1] This we can also express as being between 96 and 192 miles per second per million light-years but, whatever units we use, it gives a very large figure. It also gives something else— a new way of measuring distances of galaxies that are too far away for us to detect novae or Cepheids, because it is only necessary to obtain a red shift and then, from this and the distance-shift relationship, the distance is given. The method cannot give a precise distance, but will give the astronomer a general idea, and now we find that billions of light-years are involved when it comes to determining how far away the farthest galaxies lie.

[1] One megaparsec is 1 million parsecs; a parsec is 3.25 light-years and is so called because at this distance a star has a *par*allax of 1 *se*cond.

These very large figures are owing to the discovery in 1952 by Walter Baade at Mount Wilson and Mount Palomar Observatories that there appear to be at least two kinds, or populations, of stars, and thus of Cepheids, in spiral galaxies. Baade began his research during the late 1940's when he started to use the newly invented red-sensitive photographic plates and was able to resolve into stars some of the inner parts of the Andromeda galaxy and its companion galaxy NGC 205; previously only blue-sensitive plates had been available for astronomical photography, and they had allowed only the outer parts of Andromeda to be resolved into stars. Once Baade had resolved both types of star, he realized that a spiral galaxy such as this has mainly, but not entirely, blue stars (O and B types) in its spiral arms, and yellow and red stars in its central regions. He called these population I and population II respectively. But the Cepheids used for determining distances of galaxies had been bright O and B, population I type variables, whereas the standards they had been compared with in our galaxy were population II type Cepheids which were not so luminous. When the 200-inch telescope was completed Baade had hoped to photograph the population II Cepheids (the dimmer ones) in the central parts of Andromeda, which it was estimated the telescope would show quite clearly. In fact it did not, and Baade realized that Andromeda must be farther away than previously believed because of the confusion between population I and population II Cepheids. Its distance, as well as the distance of all other galaxies, had to be multiplied by something like two and this also affected the red shift-distance relationship, or Hubble constant.

But with such figures do we come to the edge of the universe? The recession of the galaxies brought the question to the fore, especially as the theory of relativity had previously led de Sitter and two other mathematical astronomers, A. Friedmann and the Abbé Georges Lemaître, to conclude that such a recession ought to be expected for theoretical reasons alone. Indeed Lemaître showed that Einstein's scheme of the universe must be unstable— it must either expand or contract. Friedmann came to his conclusion about an expanding universe in 1922, but the paper he published was largely ignored, and only when five years later Lemaître showed how Einstein's own computations really led to a changing universe, was interest awakened. So by the 1930's, the idea of an expanding universe was generally accepted. But where was the universe expanding to?

To try to answer this means that we must go back to the eighteenth century again and consider for a moment some mathematical work apparently quite unconnected with astronomy. During the eighteenth century the whole basis of geometry was reexamined, because it had then become clear that the ideas of the ancient Greek geometer Euclid, who lived in Alexandria during the third century B.C., needed thinking through again. Euclid's geometry, the usual geometry with which people are familiar, was a logical system based on a number of axioms (statements of fact that appeared to require no proofs as they were self-evident). For example, "things equal to the same thing are equal" and "the whole is greater than the part" were two such axioms. Then there followed a series of five postulates, which also seem obvious enough and could be proved: for example, "a straight line can be drawn from any point to any other point" and "a circle can be described with any point as center, and with a radius equal to any finite straight line drawn from the center." But Euclid's fifth and final postulate is less straightforward: "If a straight line meet two other straight lines and make with them the two interior angles on one side together less than two right angles, these two other straight lines will, if produced, meet on that side on which the angles are less than two right angles." That is, if we draw a line AB (see Figure 11–2) and let two straight lines CE and FH cross it, so

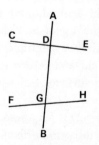

FIGURE 11–2. The problem of parallel lines. This proposition led to the development of other geometries than that of Euclid, which appeared better suited to explain events in space.

that ∠EDG and ∠HGD are less than 180° when we add them together, then CE and FH will meet at a point P if we extend them for long enough. Now, if the two angles (∠EDG and ∠HGD) are equal and add up exactly to 180°, what happens? The lines are parallel, or so it seems, but this cannot be proved to be correct under every condition. There is evidence that Euclid himself was not happy about the postulate, and later Greek geometers always felt slightly uncomfortable about it, but it is part and parcel of the scheme of geometry we use and that Euclid made into a complete mathematical system. During the 1730's an Italian priest, Girolamo Saccheri, decided that the best way to deal with this awkward postulate was to ignore it and construct all the theorems of geometry without it. Saccheri concluded this had led him to an absurd geometry and was satisfied that Euclid's was the only true one.

Between 1823 and 1829 the Hungarian James Bolyai and, quite independently, the Russian Nokolai Lobachevski, again worked out a complete system of geometry without the fifth postulate and went further than Saccheri had done. They arrived at a geometry that describes figures that cannot be drawn on a flat piece of paper—a plane—as is the case with Euclid's geometry, but can only be drawn on the surface of the trumpet-shaped figure known as a hyperbola. It is called "hyperbolic geometry." But this is not the only scheme that can be derived if we discard Euclid's fifth postulate. Another geometry was developed by a young German mathematician Bernhard Riemann in 1854. He decided on a geometry with no parallel lines—Boylai and Lobachevski had allowed there to be an infinite number—and in Riemann's geometry there is no infinitely long straight line, because if they continue far enough, all straight lines actually curve round and return to themselves. Riemann's is a geometry of curved surfaces and what we can term "curved space."

Now we must ask ourselves are these peculiar, or at least unfamiliar, geometries anything more than mathematical exercises? Could they really describe the universe? Surely Euclid was correct? Surprisingly enough, there is an answer in our own experience. Suppose we draw a triangle (see Figure 11–3a), it does not matter what sort of triangle, it has three angles inside it and these all add up to 180°; this is one of the simple results that can be proved in Euclid's geometry. But now imagine that we are going by ship from London to New York and then from New York to Belém in the north of Brazil, and back again to London. This would mean traveling in a triangle (Figure 11–3b) but, because the earth is round not flat, the path taken by our ship would be a triangle with curved lines. And this triangle would have its inside angles equal to more than 180°, as Figure 11–3b shows. So to express the mathematical way to map our earth we have to use a different kind of geometry from Euclid's—spherical geometry—where triangles can have their internal angles equal to more than 180°.

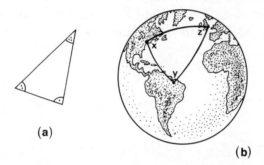

(a)

(b)

FIGURE 11–3. In a plane triangle (a) all the internal angles add up to exactly 180°; this not so for a triangle on some other surface, such as a sphere (b).

Euclid's geometry is fine when we are dealing with geometrical figures on flat surfaces, and this is all right even for use on earth when we are only dealing with very small areas such as a house or a public square in a city, but the errors it brings become too large in navigation, so spherical geometry has to be adopted. Mathematicians can express their geometries in terms of mathematical equations (we can do this for Euclid's flat surface geometry too, as René Descartes described in 1637), and do not have to draw them on paper. This is just as well, for hyperbolic geometry and Riemann's geometry are hard to illustrate in pictures.

When we apply relativity theory to the whole universe, and

come to an expanding universe such as Hubble seems to have observed, then the mathematical equations lead us to use Riemann's geometry rather than Euclid's, because the equations are simpler that way. Perhaps, then, this is an indication that space is not a sphere, nor does it extend outward forever, but folds over on itself when we come to consider the most immense distances, the kind of distances we meet with in observing the farthest galaxies. To ask how far space extends and, if it ends, what lies beyond it, is a meaningless question because we must think of the sort of space used by Euclid in order to ask it, and the universe is not Euclidean. All we can say is that the most distant galaxies are some thousands of millions of light-years away; with this the astronomers of the 1930's had to be content, and it seems that the astronomers of the 1970's will have to be content with it as well, for, at this stage, there are no more definite details of the extent of the universe.

12 ✳ Radio Astronomy

The interplay between what the astronomer observes at his telescope and what the physicist and chemist observe in the laboratory has been closely linked ever since the days of Bunsen and Kirchhoff and the development of the spectroscope (see Chapter 8). Over the years, the spectroscope has proved of immense importance in astronomy, having many more applications, and giving astronomers far more clues to the way the universe works, then even Huggins could have imagined in his most enthusiastic moments—and there has never been a keener astronomer than he. Yet this is not the only field in which the scientist in the laboratory has held the astronomer in the observatory; the detection of radio waves is another. Radio astronomy has also brought in its wake the most astounding discoveries, but it took an extremely long time before even the simplest radio observations were made. Astronomers were once again very conservative, though the lessons of astronomical history should have made them more adaptable.

The whole story really takes us back to the nineteenth century, when Michael Faraday was working at the laboratories of the Royal Institution in London. Faraday studied every aspect of electricity and, by many careful experiments, managed to see what everyone else had missed—that all electrical effects resulted from the same cause. Today this seems obvious enough, but in Faraday's time physicists considered that electricity that came from a battery and flowed along wires was quite different from the sparks and flashes of electricity generated when silk and glass were rubbed together. What is more, both were thought to have little if anything to do with magnetism. But once Alessandro Volta had invented the electric battery in 1800 and a continuous flow of electricity was at last available, a number of people, especially the Dane Hans Oersted, the Frenchman André Marie Ampère, and the German Georg Ohm, began to find links between battery electricity and magnetism. However it was left to Faraday to tie these views together with what was then called static electricity, which was generated by friction machines, and to show that all electricity was of the same kind and linked with magnetism. As he put it, "Electricity, whatever it may be, is identical in its nature," and this led him to introduce the ideas of the electric field and the magnetic field, the concept of electricity and magnetism both stretching out through space and interacting with each other.

Faraday's experimental results brought in their train many practical developments such as the dynamo and the electric motor and, on the theoretical side, they led to mathematical studies on electrical and magnetic fields, first by William Thomson (better known as Lord Kelvin), and James Clerk Maxwell. From these studies in 1861 Maxwell produced a series of mathematical equations that showed precisely how electricity and magnetism interact with each other. When they were published three years later the scientific world was amazed at their extraordinary similarity to the equations derived a few years earlier for light waves. Clearly, here was something of profound importance. In 1845 Faraday had performed an experiment showing how a light beam could be disturbed by a magnetic field, but Maxwell's discovery of just how electricity, magnetism, and light were closely connected was a breakthrough. It meant that light was really an electromagnetic wave and, if this were so, then it should in theory be possible to reflect and refract all electromagnetic waves, not only those of light. Yet it was more than twenty years before anyone was able to prove this experimentally. In 1888, at a laboratory in the polytechnic at Karlsruhe, southwest Germany, Heinrich Hertz managed to generate electromagnetic waves that could not be seen, but that he could detect with electrical instruments, and he went on to find that he could reflect these from metal surfaces in the same way light would be reflected from mirrors. These electromagnetic waves were radio waves.

After Hertz's experiments, other physicists took up the matter

and new facts were revealed. The range of wavelengths of radio
waves was found to be very extensive, from ten centimeters to at
least twenty-five kilometers (four inches to fifteen inches).[1] The
British physicist Oliver Heaviside and the American A. Kennelly
suggested the existence of an electrified layer of air some seventy-
five miles above the earth that reflected short radio waves. This
was important because it could explain how, by bouncing radio
waves off the electrified layer and off the sea, Guglielmo Marconi
had managed to transmit and receive messages across the Atlantic,
a feat that many had believed to be impossible because of the
curvature of the earth and the fact that radio waves, like light,
travel in straight lines. Yet, though Heaviside's suggestion was
made for theoretical reasons in 1900, and Marconi's transmissions in 1901, experimental proof did not come until 1925. Now it
is known that there is not one electrified layer but four, and these
affect radio astronomy by acting as reflectors to most radio waves
and thereby prevent the major part of all radio radiation from
space reaching the earth's surface.

The beginnings of radio astronomy were accidental. The Bell
Telephone Laboratories of New Jersey decided in 1931 to carefully investigate a problem that marred good short-wave radio
communication—crackling noises, or static arising from electrical
machinery, thunderstorms, and other causes. Karl Jansky, the
engineer making the investigations, developed a special directional antenna, which could be pivoted so that he could determine
the direction from which the static came (see Figure 12–1). To
his surprise he discovered that some static seemed to come from
the direction of the sun, and some from the direction of the
constellation Sagittarius, though later tests he made did not confirm the sun as a source. In 1933 a second investigation was made
of static, this time in New Jersey, because good reception of short-
wave radio signals from England was wanted. Once again Jansky
detected cosmic static. Little notice was taken of these results, and
astronomers, if they knew about them at all, did not realize their
significance. To the astronomers light was the language of the

[1] The short wavelength end has now been extended into the millimeter
region.

FIGURE 12-1. Karl Jansky with his special directional antenna with which he discovered cosmic static. Courtesy of the Ronan Picture Library.

universe, whereas radio was a man-made means of communication.

It was not an astronomer, but an American amateur radio enthusiast, Grote Reber, who pursued the matter after having learned privately that work at the California Institute of Technology had confirmed Jansky's results. He built a special directional antenna in the shape of a bowl, thirty-one feet in diameter, which gathered radio waves and focused them in much the same way as a mirror does in an optical telescope. Reber's instrument was fixed in azimuth, but could move up and down in altitude; as the earth rotated, so different parts of the sky passed in front of the antenna and could be examined. His results were interesting; they confirmed that cosmic static came from Sagittarius, but, being more sensitive to direction than Jansky's antenna, Reber's radio telescope indicated that the static was emitted from that part of the constellation where it crosses the Milky Way. Some other patches of the Milky Way were also found to be sources of static. The results were published in 1940 and 1942, but astronomers paid little attention at the time, owing partly to the results being published in a radio journal. Reber, however, did realized the importance of what he had achieved and, in 1946, after the end of World War II, his results were at last accepted by astronomers. A little work on the sun had been done during the war by G. C. Southworth in the United States, and by James Hey in England, but now radio astronomy studies began in earnest.

The new research was carried out at the Royal Radar Establishment at Malvern, England, by physicists who were experts in radio. Here Hey and Edward Appleton set to work using a considerable amount of radar equipment that had been designed for detecting enemy aircraft. (Because of the war there was a large staff of radio operators all over the country with radar sets, and it was desirable to keep them occupied until they were free to return to civilian life.) Radar used very short radio waves, emitting these as pulses and then receiving back the reflected pulse—the radar echo. By timing the interval between emission of a pulse and its return, the distance of an object was determined, because radio waves travel at a known speed—the speed of light. First, Hey and Appleton used the radar sets purely as very short-wave directional receivers (as radio telescopes), and the Sagittarius source was observed and its radiation confirmed; cosmic static was received from a source in Cygnus, the emitting area of which they measured and found to be $2°$ in diameter. They also observed the sun, and found that "big sunspots were extremely powerful ultra-short wave radio transmitters," which radioed energy far more strongly than was expected on theoretical grounds worked out from the sun's known temperature. Moreover, they found that bursts of radio emission arrived when solar flares—very hot bright patches of gas—were seen. But this was not all; on other wavelengths they discovered that the sun was always emitting radio waves. These were not so intense as the bursts just mentioned, but they were stronger than theory led one to expect and indicated some reactions that would give more knowledge about the sun's activity when they were properly understood.

As well as using the radar sets as receivers, Hey and Appleton used the wartime radar pulse method whenever meteors appeared. After some practice they found that they could detect meteors when there was nothing visible. This radar detection was possible because a meteor generates a trail of electrified air particles as it rushes headlong through the atmosphere, and such a trail gives excellent radar echoes. Ordinarily, meteors are difficult to observe, either visually or photographically, with great accuracy, for they need at least two simultaneous observers some miles apart for their heights, velocities, and directions to be obtained. Radar has obvious advantages because heights, velocities, and directions can all be measured by one observer and, in addition, micrometeorites (about the size of sand grains) are readily detectable, whereas they cannot be observed visually at all. Yet another advantage of radar is that meteors arriving in the daytime can be traced when the glare of daylight makes them invisible to the eye or to the camera.

The radio astronomy team at Malvern were soon followed by another team at the University of Manchester that established an observing station at Jodrell Bank near Macclesfield. Here, with unbounded enthusiasm, Bernard Lovell extended the meteor work of Appleton and Hey and finally proved beyond all doubt the source of meteors. It had been argued that most were interstellar, arriving from immense distances in space, but Lovell and his colleagues showed, by measuring directly a vast number of meteor velocities, that the speeds of meteors were too low to agree with this view. They concluded that at least 90 per cent of meteors were part of the solar system.

With the success of radar meteor studies, and the discoveries made using radar sets as simple radio telescopes, the astronomer was forced to realize that astronomy had a powerful new tool, a tool as powerful, perhaps, as the spectroscope. Radio astronomy had arrived, but its results so far were only a foretaste of what was to come. In England Lovell's success with meteors, and his appreciation of the future possibilities, led him to look about for money for a really large radio telescope; in 1958, after many financial and engineering difficulties, the huge radio telescope with a 250-foot diameter dish that could be turned to any part of the sky was completed at Jodrell Bank (see Figure 12–2).

Perhaps the biggest problem that early faced radio astronomers in their work was the poor resolution of radio telescopes. The power of an optical telescope to resolve distant celestial bodies into separate objects (see Chapter 7) depends on its aperture, the

200-inch reflector at Mount Palomar being able to resolve objects as close together in the sky as 0″.023. The resolving power of a radio telescope also depends on aperture, but because radio wavelengths are at least a million times longer than the wavelength of light, the resolution of a radio telescope is far less for a given aperture. For the 250-foot instrument at Jodrell Bank, when work-

FIGURE 12–2. The giant 250-foot diameter radio telescope at Jodrell Bank, Cheshire, England. The world's first really big radio telescope, it is still (1970) the largest that can be steered to point anywhere in the sky. This photograph was taken from inside the telescope's control room. Courtesy of the Ronan Picture Library.

ing on a wavelength of thirty centimeters the resolving power is only 14′; this means that even so large a dish cannot pinpoint radio sources closely enough for them to be identified by an optical telescope without any possibility of error. But there are two ways out of the difficulty. One is to reduce the wavelength so that instead of operating at thirty centimeters the telescope operates at, say, three centimeters; the resolving power would then increase to 1′.4. Unfortunately this is not entirely satis-

factory because of the need to make observations at longer wavelengths within the so-called radio window (the range of wavelengths from one centimeter to ten meters that can get through the
atmosphere). No, the radio astronomer could not be restricted to
the shortest wavelengths, and some other way of increasing resolving power had to be found.

The radio interferometer provided the answer, for here two
radio telescopes are used at once, and the resolving power increases because it depends on the distance between the two telescopes and not on their size. The basic principle of the interferometer was well known to physicists. An optical design prepared in
1890 by Albert Michelson had been used to measure the
diameters of Jupiter's satellites, while in 1920 his design had been
applied to the 100-inch at Mount Wilson for measuring the diameters of some large red giant stars. The radio astronomers now
took this up, and the first interferometer measurements on radio
sources were made in 1948 by Joseph Pawsey and colleagues in
Australia. Ingeniously adapting an optical interferometer that
used one mirror (Michelson's used two), they set up a radio
telescope at the edge of a 300-foot cliff near the entrance to
Sydney Harbor (see Figure 12–3). Radio waves came directly

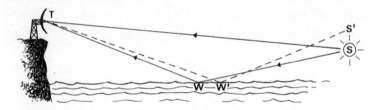

FIGURE 12–3. The arrangement devised by Joseph Pawsey in
Australia for an interferometer, using the sea as a reflector.

from the sun as it rose over the sea, S, to the radio telescope, T;
they also came from the sun to the sea, W. From here they were
reflected from the surface of the water to the radio telescope. As a
result, the radio telescope "saw" two pictures of the sun, one by
direct radio radiation and one by reflected radio radiation. The
two sets of radio waves met at the telescope and interfered with
each other. As the sun moved higher in the sky, the point on the

water at which the waves were reflected changed, and the change
caused the length of path from sea to radio telescope to alter (the
distance WT is shorter than W'T). As the length of path alters, so
the way in which the reflected waves and direct waves interfere
with each other changes—the waves in the two beams get in step
and then increasingly out of step with each other. When they are
in step the total radio energy the telescope receives is greatest (see
Figure 12–4), and this happens when the part of the sun from

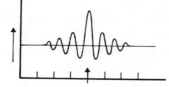

FIGURE 12–4. The signal of a celestial radio source received by an
interferometer. The use of an interferometer gives one signal with
a large peak and other smaller ones. The large peak is sharp and
lies in the direction of the source, thus making pinpointing its
position easier and more accurate than is possible with a single
antenna.

which the radio waves are coming is directly in line with the
telescope. Because of the change in radio energy either side of this
peak point, the interferometer gives far better resolution. Pawsey
found that the radio waves came from those parts of the sun where
there were large groups of spots. His resolution was 8', a figure
that another Australian radio astronomer, O. B. Slee, managed to
reduce to 1'.5 two years later.

But radio astronomers cannot always have a cliff conveniently
near, and even if they could, the method is unsuitable for observing
objects high in the sky. So at Cambridge, England, Martin Ryle
and his team began to investigate the whole question of interferom-
eter-type radio telescopes. In 1948 Ryle and Graham Smith,
using two separate radio telescopes and allowing the radio waves
they received to interfere, obtained a resolution of 8' of a night
sky object in the constellation Cassiopeia. This was promising,
and Ryle later designed radio telescopes specially adapted to
radio techniques, rather than translating optical methods into

radio terms. This may seem an obvious thing to do, to design a telescope especially for the kind of wavelengths with which it is going to work, but all pioneering efforts seem obvious after they have been successful. During the early 1950's it was a breakthrough that was to have the most important results.

One of the first developments was the realization that you did not need to have both radio telescopes positioned in the same way. In fact, if you had two long narrow ones at right angles to each other, this could give a very effective result, and in Australia Bernard Mills designed and built an instrument along these lines. Known as the Mill's Cross, the two long sets of radio telescope antennas overlap each other, resulting in a giant cross. At the long radio astronomy wavelengths, such as 3.7 meters, a resolution of 50' was obtained, and though this may sound very poor it compares favorably with 29°, which is what a single antenna such as the 250-foot Jodrell Bank instrument would give at this wavelength. But the Jodrell Bank telescope is designed to work at far shorter wavelengths and so in practice its resolution is only a few minutes of arc, and with its large bowl it is a very powerful receiver. All the same, the figures show the principle clearly enough, and make it obvious that for longer radio wavelengths an interferometer is essential.

Probably the most significant extension of the interferometer is Ryle's development, the aperture synthesis radio telescope. There are at least two antennas, one of which is fixed and the other movable. To see how it works it will be best if we look at his first design. Here there was one long antenna that stretched out in an east-west direction for over a quarter of a mile, and another shorter antenna that lay in a north-south direction (Figure 12–5), and could move in an east-west direction for 1,000 feet along a special railroad track. Observations were made one day with the movable antenna in one position—at the west end of its track, say—and then the next day at the next point along, and so on until it had arrived at the east end. The resulting observations gave the equivalent resolution of a radio telescope with the aperture shown by the dotted line. Such an instrument is not so sensitive as an equally large single antenna, but with such an interferometer one can pinpoint radio objects for study by

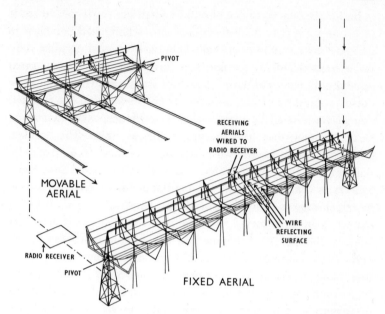

FIGURE 12–5. The aperture synthesis radio telescope interfer-ometer devised by Ryle. By moving one antenna along a railroad track, and observing with this antenna in different positions, he ob-tained an accuracy of position determination of radio sources equiv-alent to the use of a single antenna as large as the whole area bounded by the fixed antenna and the railroad track. From the *Geographical Magazine*, London.

optical telescopes and other equipment. This sort of radio tele-scope is known technically as an unfilled aperture.

Ryle has extended the method and has constructed what is known as a filled aperture. Here one antenna moves so that it fills up all the space in between two fixed antennas. Again, observa-tions must be made night after night, or day after day—a radio telescope is equally effective whether it is observing in daylight or at night. The rotation of the earth allows the telescope to sweep out an area of the sky, as it does in the case of the unfilled aperture telescope, and the observations are added together to give the final results. The addition is not simple and, in practice, the observations are automatically put on punched tape and fed to a computer. At his Cambridge observatory Ryle has the equiva-lent of a radio telescope that is one mile in diameter.

Radio astronomers have also developed their own special radar receivers to operate with their antennas. Their need has been to design receivers powerful enough to pick up and amplify faint radio signals, but in so doing they have been up against a great difficulty. Usually the more powerful a radio receiver is, the more noise—a kind of hissing sound—it generates inside itself. But if you play radio waves from space over a loudspeaker instead of charting them, they also produce a hissing noise. The problem facing the radio engineer is to reduce the hiss from the radio receiver so that it does not mask the hiss from the radio sources in space. One way of doing this is to construct a receiver in which the incoming radio waves from space stimulate, or trigger, a burst of waves from the receiver. Because it uses very short radio waves, or microwaves, a receiver that does this is known as a *m*icrowave *a*mplifier by *s*timulated *e*mission of *r*adiation, maser for short. To work properly such a radio receiver must be surrounded by liquid helium gas to cool it, but in spite of this disadvantage it is one of the most efficient receivers developed since the late 1950's. The principle on which it works has also been used at visual wavelengths and is known as the *l*ight *a*mplifier by *s*timulated *e*mission of *r*adiation, or laser; it has been used with a radar technique to bounce light waves back from the moon. Another kind of radio receiver uses radio waves from space to cause a burst of radio emission in a different way—by continually altering one of the component parts, or parameter, of the receiver. This is known as a parametric amplifier.

This sketch of the development of a radio astronomer's observing equipment is important in any historical account of the subject because, without it, none of the surprising discoveries could have been made. The increase in resolving power since 1946 has made it possible to identify more and more of the objects that are giving out radio radiation. If their nature is to be understood, astronomers need to be able to obtain some kind of visual picture; after all, it is in visible light that their picture of the universe has been built up over all the previous centuries. As has happened before in the history of astronomy, fresh understanding of the universe has had to wait for new ways of observing it.

It is not possible in this book to go into detail about the results that radio astronomers have obtained. Instead, we must content ourselves with a general sketch of what has happened, so that the way radio observations have affected the astronomers' whole outlook on the universe may be gauged; this is its historical importance. As a beginning it will be better to look at how the radar work of Hey, Appleton, and Lovell has been followed up.

Radar is limited in use because a radio pulse shot out and reflected back loses energy on its journey. How far it can go from the radio engineer's point of view depends on how much energy he can pack into each pulse and how large an antenna and how sensitive a radio receiver he can build to pick up the echo. But this is not all. Radio pulses travel with the velocity of light, so there is an astronomical limit, too. The nearest star lies at a distance of four and one-third light-years, so a radar pulse to it is going to take eight years and eight months before it returns. To go out very much farther, even if the radio engineer could manage it, would give an inconveniently long wait between pulse and echo, but to the sun and planets the interval amounts only to minutes or hours.

For this reason radar observations have been confined to meteors and to the sun and planets. By the time the giant radio telescope was built at Jodrell Bank in 1958, and efficient radio telescopes were brought into commission in the United States, radar work on meteors had largely given place to attempts to receive echoes from the moon, Venus, and the sun. The success of these results had led to new determinations of the distances of the sun and moon, and in 1968 the Massachusetts Institute of Technology used radar for mapping the surfaces of the moon and nearby planets. Another investigation made with the 1,000-foot diameter radio telescope at Arecibo, Puerto Rico, in 1953, measured the wavelength of echoes returned from Mercury and Venus. (See Figure 12–6.) Because the planets rotate, echoes from the edge moving toward the observer will appear to have a shorter wavelength than those reflected from the edge that is moving away. These observations have given a rotation period of fifty-nine days for Mercury, in place of the eighty-eight-day period previously accepted, and 243 days for Venus, which has been found to spin around from east to west instead of moving in the west to east direction of the other planets.

But the most important observations, which will go down in history as opening a new era in astronomy, have been those using purely the radio signals received from bodies in space. At a

FIGURE 12–6. Built in a giant natural bowl in the ground, the diameter of the radio telescope at Arecibo, Puerto Rico, is 1,000 feet. It cannot be steered more than a few degrees, but different parts of the sky can be observed because the telescope is carried around by the earth's rotation. Courtesy of Cornell University.

number of radio astronomy observatories, maps of the sky showing the positions of radio sources were made from 1951 onward, the first big survey being carried out at Cambridge, England. Some 100 sources were charted and, using an interferometer, Graham Smith was able to determine the positions of two of the most powerful with an accuracy of 45″ in one direction, and no more than 10″ in the other. One source lay in the constellation of Cygnus and the other in Cassiopeia. These positions were sufficiently accurate to allow a telescope to look at the points in the sky where these powerful radio sources were, but nothing special could be found with small telescopes. However, the next year Walter Baade used the 200-inch telescope. The Cassiopeia source was found to be the remains of a supernova lying in our galaxy, but the Cygnus source was more puzzling. It appeared to be two spiral galaxies in collision though later studies have shown that, far from colliding, the galaxies may be moving apart. The results were not only exciting, but also made radio astronomers, if no one else, aware that radio telescopes were powerful aids to ordinary astronomy; at Cambridge, Martin Ryle began to design bigger and better interferometers, culminating in his aperture synthesis instrument. Some of the results of this will be considered in a moment, but to keep our story in its historic sequence, we must look at other work first.

During World War II a few astronomers were able to remain at Leiden Observatory in Holland, and because they were prevented from making any observations they kept themselves busy with astronomical theory. One of these was Henrik van de Hulst who, by applying the theory of the way in which atoms emit radiation, concluded that hydrogen gas not hot enough to glow and emit light should give out radio waves at twenty-one centimeters when its electrons spun as they orbited the nucleus, and its atoms collided. In the space between the stars in our galaxy there should be enough collisions for detectable radio waves to be emitted. Soon after the end of the war, physicists in the United States found it possible to make hydrogen do this in the laboratory; they measured the radio wavelength and found it to be twenty-one centimeters. Experiments to build radio telescopes to detect this radiation met with great difficulty, and though Van de Hulst came to his theoretical conclusion in 1945, a conclusion independently arrived at two years later by the Russian astronomer I. S. Shklovsky, the first observations of it were not made outside the laboratory until 1951.

The importance of being able to observe hydrogen gas that is not glowing is hard to exaggerate. With an optical telescope it is impossible to tell it is there, except when a cloud lies in front of part of a starry background and obscures some of the stars. In those sections of the sky far from the Milky Way, radio is the only means of detection; because astronomers want to discover the nature of our galaxy, precisely where this gas lies must be found. A great deal of time has been expended on this search and on searches for other gases that give no visible effects but can be detected by radio telescopes, with the result that everyone had a far better idea now of how much gas there is between the stars. But the most exciting discovery made using radio studies of the dark hydrogen in the galaxy has been that the galaxy is a spiral. There were clues that made this seem likely, clues that came from studies of the motion of the stars and the way in which they are distributed in space, but no one could be certain. Once it became possible to pick out clouds of hydrogen, it was found that they clearly showed that there were a number of spiral arms. The northern skies were observed by radio astronomers in the Netherlands, while radio astronomers in Australia dealt with the southern skies. Though it was still not possible to observe the central regions of the galaxy, by 1958 there was no doubt about what kind of galaxy ours is.

Other studies of the galaxy, using radio telescopes, have been made and these have shown that even though no one can be sure of all the details about what lies at its center, there is certainly some nonglowing hydrogen nearby and the area behaves as if it is a very strong magnet. At the center itself there is some electrified hydrogen and probably other material as well. A further interesting fact that has come out of observations made at the National Radio Astronomy Observatory at West Bank, Virginia, is that there is some ammonia vapor and formaldehyde vapor in space among the stars. This was only found during 1968 and 1969, so it is too early to do much more than report the fact and point out that both substances play an important part in the formation of living matter; the real significance of their presence can be appreciated when one considers the possibility of life elsewhere in the universe.

Besides clouds of gas, from 1950 onward radio astronomers have detected radio waves coming from very small areas of the galaxy. Some of these are caused by bright gas clouds, and some by the gases still rushing outward from the remains of a supernova. Others come from flare stars, that is, small red stars that show sudden outbursts of light that last for only a few minutes. But the most spectacular discovery about the inside of our galaxy was made by Jocelyn Bell and Anthony Hewish at Cambridge, England, in 1967. They found bodies that emit pulses that are repeated every few seconds. The pulses themselves are very brief, lasting no more than about .04 second each, but the objects that give rise to them are a mystery and are still being argued about. However, when they were discovered there recently had been a great deal of discussion on the question of whether there are other intelligent creatures in our galaxy and, if there are, how they would communicate with us. Were the pulses received at Cambridge caused by messages from another civilization? For a time the Cambridge radio astronomers jokingly referred to the pulsing radio sources as LGM's—*little green men*—but even if they were amused with the idea, this did not stop them testing it thoroughly. The pulses, some more powerful than others, were studied using a computer, but no pattern could be detected. It seemed, then, that these objects, or pulsars, must be stars. In 1969 John Cocke and his colleagues at Steward Observatory, Tucson, observed one pulsar visually, and a photograph from Lick Observatory has shown that this pulsar is one of the stars that can be seen in the crab nebula that is the remains of a supernova that occurred in 1054 A.D. The nature of the pulsing star is uncertain, though it may be a neutron star—a star that has come close to the end of its life and has shrunk to no more than about ten miles across. In the middle of such a star the centers of its atoms would be crushed together so fantastically that the only parts remaining would be neutrons, and a piece the size of a sugar lump would weigh 25 million tons. Thomas Gold has suggested that a neutron star spins round very quickly, and should have a strong magnetic field, so that when gas oozes out and is carried around by the very high rotation it will emit radio waves. Whatever the answer, this explanation shows how the facts of nuclear physics, magnetism, and the behavior of radiation are now all part of the astronomer's world.

Radio astronomy observations beyond the galaxy have brought new facts, too, some of them very startling. Examinations have been made of nearby galaxies, such as the one in Andromeda, which lies some 2 million light-years away; so far they have been shown to be similar in every way, with objects in them radiating radio waves just as the gas clouds in the galaxy do, and presumably they have pulsars as well. But the greatest number of radio objects lying beyond the galaxy are what have become known recently as radio galaxies. When first detected and identified in 1951, they were called peculiar galaxies because as each radio source was photographed it showed a galaxy with peculiar properties. Eventually, some turned out to be elliptical galaxies emitting a jet of material out into space, whereas others were ellipticals with at least two separate areas from which radio waves were coming. A small number have been identified as Seyfert galaxies, which are the small galaxies with bright erupting centers first discovered optically by Carl Seyfert in 1943.

These observations of galaxies that are undergoing explosions or have suffered explosion in the past and, in a few instances, seem either to be galaxies colliding or splitting, are bound to play a vital role in our ideas of the universe; in fact they are already doing so, as we shall see in Chapter 15. But radio galaxies are not the only extragalactic radio sources to be observed. There is another kind, and their discovery is one of the most dramatic stories in the whole of modern scientific research.

From 1950 onward, Martin Ryle and his colleagues have been making radio surveys of the sky. Their lists of radio sources are numbered 1C for the first Cambridge catalog, 2C for the one made between 1953 and 1954, and so on. Australian radio astronomers have also made surveys, and in 1957 David Dewhirst of the Cambridge University Observatories took the newly completed 3C catalog and one from Australia, to Mount Palomar, and examined the original photographic plates of the Mount Palomar sky survey made during the 1950's. After a year's work, Dewhirst found that of the ninty-three brightest sources in the catalogs, which contained a total of 450 objects, only twenty-one could be definitely identified with visual objects. However, in 1960, John Bolton and J. Matthews, two radio astronomers working at the California Institute of Technology, identified one object in the 3C catalog with a dim blue star. Known as 3C 48, this aroused little excitement, though it was admitted that it was unusual as it seemed to be a star rather than a hazy patch of light as was normal with a radio source. They took it to be a star in our galaxy. A few other identifications were made, and all the objects seemed to be hot blue stars that had some mildly surprising features, the most important of which was the unusually great strength of radio waves that they were emitting. (See Figure 12–7.) Two years later

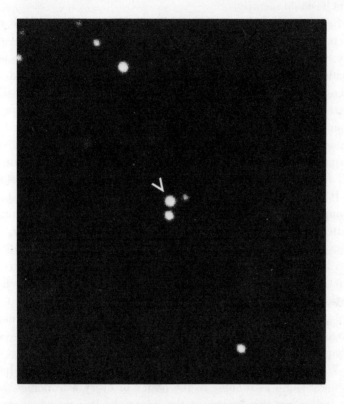

FIGURE 12–7. Original photograph of quasar 3C 147. The quasar is the dot at the point of the arrowhead, and visually looks just like another star. Courtesy of the California Institute of Technology.

Cyril Hazard in Australia observed 3C 273, taking advantage of being able to watch it with a radio telescope while the moon passed in front. This ingenious way of studying the source meant that he was able to fix its position very accurately, because the moon's place in the sky is known very precisely. Hazard discovered that 3C 273 was really a double radio source, and his observation of its position allowed astronomers at Mount Palomar to observe it in detail optically. John Oke found that the object emitted heat waves very strongly, and Maarten Schmidt, a Dutch astronomer, was surprised when he examined its spectrum because, though he observed a great many lines, there was not one that he could identify.

Schmidt puzzled over the problem, and decided to publish the facts in a scientific journal so that others could think about the mysterious lines. This was in February 1963 and, as he was writing the paper, he suddenly realized that the lines he could not recognize formed the pattern characteristic of hydrogen. But they were all in the wrong place, each having a wavelength far longer than usual. In other words all the lines in the spectrum showed an immense red shift, much larger than anyone expected. A colleague, Jesse Greenstein, immediately checked on the spectrum of 3C 48, and found that it showed an even more fantastic red shift.

It has now been shown that these objects are certainly not stars, though they look like them at first glance. For want of something better they have been named quasistellar radio sources, or quasars, but their nature has still to be explained. Since 1963 it has been discovered that they show bright lines in the spectrum, but the sort of lines to be seen in a glowing cloud of gas, as well as dark ones, and some have more than one red shift. Possibly they are galaxies very far away in the universe, perhaps in a very early stage of their lives, though this is not the only explanation (see Chapter 15 for a further discussion of them).

Whatever quasars turn out to be, it is certain that they have brought radio astronomy to the notice of every astronomer, and no longer can this branch of the science be thought of as something for the specialists only. Even though it is really only twenty-five years old, the radio telescope is now as much one of the astronomer's tools as are an optical telescope and a camera. This means that radio and optical astronomers are working in closer and closer cooperation, and their ideas are interacting to produce an even grander and more interesting picture of the universe.

13 ❋ Space Probes and Astronomy

On October 4, 1957, the Russians launched the first artificial satellite—Sputnik I. It moved in an elliptical orbit that took it as high as 587 miles and as low as 135 miles. As it orbited the earth, it sent out radio signals, which were received on the ground, varying as they traveled through less or more atmosphere, depending on how high the satellite was. Merely by receiving the radio signals, it was possible to learn new facts about the upper atmosphere; the scientific use of space probes to give information unobtainable in any other way was demonstrated beyond all doubt. Yet there had been plenty of scientific men who ruled out the idea of launching a space probe as nothing more than fantasy, and one eminent modern astronomer who, when asked what he thought of the idea of space-age astronomy, even replied that it was "utter bilge." But then there were scientists years before who had believed the airplane to be an impossible dream because they felt sure that a machine heavier than air could never fly. And so it has always been; men who are clever and knowledgeable, but who lack that vital spark of imagination, severely criticize their more far-sighted and visionary colleagues.

The idea of conquering space goes back a long way. In the second century A.D. the author and lecturer Lucian of Samosata wrote *True Stories* and described a sailing ship that is caught by a whirlwind and carried to the moon. Though he admits that, in spite of the title, the stories are fiction, his description of the lunar inhabitants that the mariners find has been a model for many later authors. Kepler, too, wrote a famous science fiction book in 1609. Called *Somnium* (*The Dream*), and published after his death, it rejected the idea of any air or other substance lying between the earth and moon, so that Kepler could have no ships or animals carrying his voyager aloft. He solved the problem of transport by having his travelers carried by demons. In contrast to this flight of fancy, his description of the moon was based on the known scientific facts, even though he did populate it with creatures that lived in cracks and crevices in the rocks. It seems that Kepler originally had the idea of describing a moon journey some years earlier, when he was trying to counter arguments against the suggestion that the earth moves in space; he thought he could build a case by describing the earth's behavior as a lunar observer would see it.

Others followed with more ideas about trips to the moon, and during the nineteenth century Jules Verne launched his astronauts with a large gun. And not the moon only, but Mars and other planets were the science-fiction writer's ports of call. Since the 1920's rockets were the favorite means of transport across space, a method that had good scientific reasons behind it. A rocket derives the energy to drive a vehicle forward from the thrust of its escaping gases on the rocket casing, and not because they push on the surrounding air. The principles of such action were explained by Isaac Newton in the *Principia*, though it was to be two centuries before they were applied to the problem of space exploration. All the same, rockets and steam jets for driving things were known very early, though steam jets were not applied seriously by the Greeks or the Romans. Rockets were probably invented in China, for the Chinese discovered gunpowder, sometime in the eighth century A.D., and gunpowder was necessary before the first rocket could be made. They were used for two purposes, first as fireworks at celebrations and religious festivals, and very soon as weapons. Strangely enough the first space rockets were to be used as weapons in this century, before they were applied to peaceful and scientific ends.

Of the engineers who tried to devise rockets that would travel a long way, and do so efficiently, the first to realize their true abilities were the Russian Konstantin Tsiolkovsky, the American Robert Goddard, and the Hungarian-born German Hermann Oberth. Tsiolkovsky was a schoolteacher, and in his spare time investigated the whole subject of rocket propulsion in space; in 1883 he had come to the conclusion that a rocket was the only device that could be used, and by the end of the nineteenth century he had worked out the mathematical theory of rocket propulsion, including the speed necessary to escape from the earth's gravitation. Tsiolkovsky even went so far as to propose the use of liquid gases as fuels and to suggest the idea of a multistage rocket, in which each stage dropped off after it had fired and used its fuel—the technique adopted today. Tsiolkovsky was born in 1857 and died in 1935, and though he was honored in the Soviet Union, no practical steps were taken to follow up his ideas.

Robert Goddard, who has been called the Father of Modern Rocketry, was born in 1882 and died in 1945. Like Tsiolkovsky, he believed in multistaged rockets, but he was a man of great practical ability and was not content only to investigate the theory of rocket propulsion, but while professor of physics at Clark University, Worcester, he began testing rockets powered by solid fuel. Later he developed liquid fuel rockets and launched his first at Worcester in 1926. By 1930 he had had some real success, and by 1938 at Roswell in New Mexico a rocket reached a height of almost 5,000 feet—nearly a mile. This is nothing compared with the single-stage rockets that now regularly reach more than 100 miles, but it was a beginning, and encouraged Goddard to look for more powerful liquid fuel drives. But it was in Germany that the liquid fuel rocket was finally developed, and then not for space exploration but for military use. In 1923 H. Oberth published *The Rocket into Planetary Space*, which in his own country drew attention to the possibilities of rocket propulsion, and in 1929 he took matters further with *The Road to Space Travel*. Goddard had avoided publicity, but Oberth welcomed it. Like Goddard, though, he was hard put to raise funds for his experiments; yet in spite of setbacks interest was aroused and rocket societies grew up in the United States, Russia, and Europe during the 1920's and experiments were carried out with increasing success. In Germany, in 1928, two rocket-driven cars were built, one of which reached a speed of 125 miles per hour on a race track, and a rocket-driven glider and a rocket-powered airplane were also made. After a lapse in general interest, Oberth's efforts seem to have reaped a rich reward, even if not in quite the way he hoped; in 1932 the German army started to take a serious interest in rockets as weapons. They were not alone, for Britain, Japan, the Soviet Union, and the United States all had rocket equipment during World War II. But it was in Germany that big rockets were developed under the technical direction of Wernher von Braun, culminating in the V-2 which was more than forty-six feet long, weighed more than twelve tons, had a range of 200 miles, and was powered by liquid fuel. The immense problems of designing a chamber inside the rocket where the liquid fuels could mix and burn, and on which the resulting gases could exert their powerful thrust, but which at the same time would not overheat, had at last been solved.

After World War II, with von Braun working in the United States, the development of the V-2 rocket went ahead, aimed now at producing powerful defense missiles: in the Soviet Union others from the German V-2 development plant went to work on a similar task. But in both cases the scientists concerned wanted to develop rockets for a space program, and the launching of Sputnik I in 1957 was a public display of what could be achieved. For astronomy, both scientifically and historically, this launching was one of the heralds of a new era of immense importance.

Now it had become possible to launch rockets into space, the astronomer was able to observe those extreme ranges beyond the visible ends of the spectrum that could be examined only from above the atmosphere. Strangely enough his studies into parts of the invisible spectrum began in 1800, when rockets and space probes were still a wild dream. Again it was William Herschel who, wanting to observe the sun, began investigating the effects of various forms of colored glass to reduce the glare and heat of sunlight. He discovered that most of the heating effects came from rays that were invisible to the eye, but could be detected by a thermometer. (See Figure 13–1.) These infrared rays lay beyond the red end of the spectrum. In 1801 Johann Ritter found that rays beyond the violet end of the spectrum—ultraviolet rays—existed, which gave no heat but did blacken a light-sensitive material such as silver chloride.

The difficulty during the early 1800's in finding an explanation for these rays was that no one then had a really clear idea of the nature of light and, so far as infrared was concerned, no correct

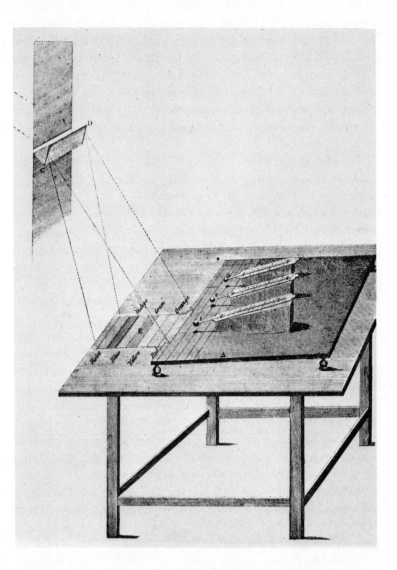

FIGURE 13–1. William Herschel's experiment using thermometers
and a prism to prove that radiation which had a heating effect but
was invisible to the eye, lay below the red end of the sun's spectrum. This became known as infrared radiation. From *Philosophical Transactions*, 1800. Courtesy of the Ronan Picture Library.

theory of the nature of heat either. Yet within the next twenty years, a new wave theory of light was advanced by Augustin Fresnel and Thomas Young, and by 1840 theories about heat as a form of energy were becoming current. Gradually the ultraviolet and infrared were understood to be invisible extensions of the visible spectrum. However, using thermometers to detect infrared was not satisfactory, and though ultraviolet could be detected photographically, this technique proved to be limited because of the earth's atmosphere. Only gradually did new and more efficient apparatus become available.

The first means of detecting and measuring infrared radiation accurately came from Thomas Seebeck's discovery of thermo-electricity in 1822. Seebeck found that if he had an electric circuit made up of at least two different metals and kept one of the places where the metals joined together hotter than the other junction, an electric current was created. The amount of current depended on the difference in temperature between the joints and on the kinds of metal used. This result was put to work in the thermopile, a device in which dissimilar metals are used for measuring tempera- ture; in 1846 this was used by astronomers to measure the heat of moonlight. The thermopile, or thermocouple, was not so sensitive an instrument as could be wished. As usual, the astronomer wanted a device that surpassed the capabilities of what was readily available, but it was some time before the thermocouple could be superseded. In 1880 Samuel Langley, director of the Allegheny Observatory, Pittsburgh, developed the bolometer, or beam (of infrared) measurer. This consisted of a thin platinum wire coated with lampblack, which absorbed the radiation falling on it. A small electric current was passed through the wire and thence to a sensitive meter. When heat fell on the wire, its electri- cal resistance changed and the reading on the meter altered. With his bolometer, Langley was able to prove that the moon radiated its own heat as well as reflecting heat from the sun. The device has been used in countless observatories ever since. In fact Langley's bolometer, with minor modifications, had remained virtually standard equipment until the recent development of such elec- tronic devices as the lead sulfide cell and such special transistor devices as germanium detectors.

The earliest large-scale surveys of very cool stars to be detected by their infrared radiation were made in 1937 by Charles Hetzler at Yerkes. In 1965 special photographic plates were used in Mexico. The trouble in both cases was that the photographic observers were limited to infrared waves no longer than .00001 inch; inasmuch as infrared rays can be as long as .004 inch, they were only observing the edge of the range. There was a vast amount of exciting information in space that could not be grasped. Electronic methods have altered the picture somewhat, and successful observations have been made of very cool stars that emit no light, but whose infrared radiation shows their temperatures to be about ten times less than the sun. The knowledge that such stars exist is important and may well cause the revision of some current ideas of how stars evolve. But for a full understanding of stellar evolution we really need to be able to cover the whole infrared range, right down to wavelengths as long as .004 inch. The trouble is that for wavelengths longer than .00016 inch, the earth's atmosphere acts as a very effective filter, soaking up every bit of the long wave infrared. To observe this range, then, it is necessary to use detectors mounted in rockets or in rocket-launched space probes.

The same problem exists for the other end of the spectrum, and though photographic observations of the ultraviolet have been made since the 1890's, they have been very limited. Here the earth's atmosphere absorbs all wavelengths shorter than .0001 inch and for a full analysis the astronomer needs to go down to .0000006 inch, and even shorter when he goes into the x-ray and gamma ray ranges. Though during the 1930's it was realized that there must be advantages in looking at the universe through x-ray eyes, no one knew just how important such observations could be, and as there was no practical way of making them, there seemed no sense in worrying about them.

Four main factors have acted to change the situation. The first of these, the development of the rocket, we have already seen. This made the practical difficulties no longer impossible to overcome, even though conquering them might be expensive. The second has been the astounding success of radio astronomy. Starting from chance beginnings, and beset by a host of doubts, it is no exaggeration to say that it has revolutionized astronomy. This immense success has led astronomers to wonder what further revolutionary results may be in store using other ranges of invisible astronomy; progress with a little of the infrared has shown something of this. Third, there has been the combining of astronomy with physics. Spectroscopy began this alliance of disciplines at the turn of the century, but the discovery of atomic power, the study of nuclear physics, and the consequence of Einstein's theory of relativity have intensified this bond. The fourth factor arises because there are far more scientists in the world now than there were prior to 1940. Now, with an increasing world population, more people are trained as scientists and are in government and private employ; of these, some are engaged in pure research, whereas others work on defense programs. In space astronomy the two meet.

The first space probes to be launched were sounding rockets, single-stage rockets that carried scientific instruments to heights of about 100 miles. Their advantages were that the rockets were simple and inexpensive. What is more, the instruments sent up in them could be recovered after they had fallen back to earth, slowed down by a parachute. This meant that they could be fairly straightforward, without the extra electronic equipment required to send the observations back to earth, and with no need for elaborate tracking stations on the ground.

Many nations have launched sounding rockets, the United States, toward the end of the war, sending up small WAC Corporal rockets to heights of forty miles. But research using rockets really began in earnest about late 1949, when the Russians launched geophysical rockets, which was followed in 1950 by the launching of the American Viking series. Since then hundreds of rockets have been launched and have been providing good and useful results, though this method may well become out of date by the 1980's.

Since 1950 the sounding rocket technique has been used to study the sun in very short wavelengths, particularly in the short ultraviolet and x-ray region. The highly energetic radiation that gives ultraviolet and x-rays has been found to come from the corona, the upper region of the sun's atmosphere, which can be

seen as a pearly colored light round the sun during the few
minutes of a total eclipse. The areas from which these radiations
emanate lie above sunspots, and the evidence obtained will help in
unraveling the questions of how sunspots form and why more
appear every eleven years. (See Figure 13–2.) Ultraviolet studies,

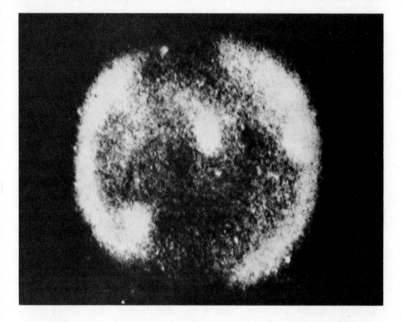

FIGURE 13–2. A view of the sun taken only by the x-rays it emits
(the bright areas are those from which the x-rays are coming).
Photographed in August 1965 from a British Skylark rocket launched
from Woomera in Australia, this was one of the first detailed x-ray
photographs from a stabilized rocket. Courtesy Science Research
Council (U.K.), Astrophysics Research Unit, Culham Laboratory,
and Dr. K. Pounds, University of Leicester.

especially those made with a spectroscope in the rocket, have
enabled details of the sun's permanent higher energy radiation to
be obtained, and these are beginning to provide further clues
about the way energy is generated inside the sun, and therefore in
other stars. This generation is known to result from the breakdown and build up of atoms—thermonuclear reactions—but details of the processes are still needed.

Some glimpses of x-ray stars have also been obtained, and enough information has come the astronomer's way to show that ultraviolet radiation would be likely to bring many surprises. It has therefore been his ambition to have space probes launched that could be devoted entirely to astronomy. These would necessarily be expensive, because not only must they be launched by multistage rockets, but they also require such observing equipment as telescopes and spectroscopes, recording equipment, and radio receivers and transmitters. And though it may seem a little strange to be continually referring to cost, this has been, and still is, an important consideration. No modern history of science can ignore this factor because it is one of those outside influences that affect what is done. In the mid-twentieth century, private funds cannot cover most research. Now the scientist cannot always achieve the results he wants, and which he knows are attainable, because the costs involved are so high.

All the same, since Sputnik I, many space probes have been launched carrying a variety of experiments. In 1962 one of these even had a radio telescope on board, and in the same year the first wholly astronomical probe was launched. The Orbiting Solar Observatory 1 (OSO 1) was to observe the sun's ultraviolet spectrum and to make some x-ray observations. In 1965 another OSO was put up, and in 1967 OSO 3 and OSO 4 were launched. Orbiting at an average height of 350 miles, the OSO's are powered by solar cells—flat plates that receive sunlight that they convert into electricity. On board are spectroscopes and telescopes, and various special pieces of experimental equipment to give answers to particular questions. Astronomers feel that the results obtained to date have made the projects well worthwhile; these probes are still sending back information that is of importance and provides a host of details about the short wavelength and of the solar spectrum as the eleven-year sunspot cycle alters.

As far as ultra-short wavelengths outside the solar system go, the first attempt to achieve something as good in its way as the orbiting solar observatory was made in 1966 with the launch of the first Orbiting Astronomical Observatory, OAO 1. The launch was successful enough, but unfortunately there was a power fault soon after and no observations could be transmitted back to earth;

the satellite is useless. This was a bitter disappointment, because shortly before the Russian satellite Cosmos 51 had made the first measures of very short ultraviolet stellar radiation and this, together with the results from sounding rockets, such as the Aerobee-Hi, and from pictures of ultraviolet star spectra taken with hand-held cameras by astronauts in Gemini 10 and 11, had shown

FIGURE 13–3. An orbiting astronomical observatory, remote controlled from earth. It makes observations of the stars and other celestial objects in radiation wavelengths that never reach the earth. Courtesy of the National Aeronautics and Space Administration.

exciting possibilities. The second OAO was not launched until December 1968, but its results have already proved valuable, as might be expected in view of the immense complexity of equipment that it carries. (See Figure 13–3.) On board OAO 2 are eleven telescopes, four with an eight-inch aperture, four with a twelve-inch aperture, and three with a sixteen-inch aperture. Each telescope contains its own special observing equipment, designed

to work in the very short ultraviolet. The four eight-inch telescopes and the three sixteen-inch ones are designed to measure the
brightness of the ultraviolet spectra of stars, nebulae, and galaxies, whereas the four twelve-inch instruments are mapping the
entire sky in ultraviolet wavelengths. Within months of launching
many of the stars observed by OAO 2 have been found to be
anything between six to forty times stronger in the short ultraviolet they emit than astronomers expected, and the nearest spiral
galaxy to ours, the big spiral in Andromeda, has been shown to
have some very hot stars in its central regions, again a result that
was not anticipated. Clearly the use of such devices as OAO 2 is
going to change astronomy radically, because it means that the
absorbing blanket of the earth's atmosphere is no longer a barrier.

It is within the solar system, though, that space probes have
had their most spectacular successes. The moon was the first body
to receive attention, a natural choice in that it is the nearest
celestial body to the earth. The earliest attempt to explore it was
the launching in August 1958 of a Pioneer spacecraft, but it was a
failure. The next Pioneer, launched two months later, managed to
get 70,700 miles from the earth, and again failed. Two other
Pioneers were launched in the two subsequent months and were
also unsuccessful. At this time, the Russians were attempting to
survey the moon, and in January 1959 Lunik I approached to
within 4,660 miles, and seven months later Lunik II crash-landed
on it. In these early lunar experiments the most spectacular feat
was the photographing of the far side of the moon by Lunik III in
October 1959. For the first time man had information about that
face of the moon perpetually turned away from the earth. It was
found to have far fewer seas, or maria, than the side we can see
and to be more rugged, facts that have a bearing on theories of the
moon's formation and the cause of its many craters. Of even
greater significance, Lunik III brought home to a wide public,
which had no particular interest in astronomy, what could be done
with a space probe. This was important, certainly in the United
States, because the public, through their elected representatives,
would have to agree to pay for any further space research.

In the years that followed, both the United States and Russia
continued to send up lunar probes. It is impossible to tell how

many failures the Russians had, because they do not carry out the launches in the same blaze of publicity as the United States; the giant 250-foot radio telescope at Jodrell Bank has picked up transmissions that such probes send out, and sometimes precipitated a Russian announcement of an unsuccessful launching. This is how we know that between April 1963 and December 1965 the Russians made at least five attempts to soft-land on the moon. Of course such failures are only to be expected in a field bristling with novel technical problems, and when in February 1966 Luna IX soft-landed on the moon and sent back close-up pictures of the surface, it was realized that another step forward had been taken in space exploration.

Meanwhile, after a series of failures, the United States had some success with their Ranger spacecraft launched in 1964 and 1965 and then, at the end of May 1966, they achieved remarkable results with the soft-landing of Surveyor 1. As well as this, there was a series of five successful low flying orbits of the moon between August 1966 and August 1967 using lunar Orbiter space probes that sent back pictures of the lunar surface. Both sides of the moon were examined and details down to one meter in size detected; indeed the probes were even more successful than had been hoped. The next milestone in lunar exploration was the outstanding achievement of a manned landing on July 21, 1969. From an astronomical point of view this meant that a number of experiments could be carried out that were not possible before. First, samples of lunar rock and surface material were brought back to be analyzed in the laboratory, and these will yield information that will help solve the question of the moon's origin and the subsequent formation of its features. Second, a seismometer was left on the surface to provide evidence of lunar earthquake shocks and of landslides caused by the heating and cooling of the surface rocks. (See Figure 13–4.) Third, a flat mirror left on the moon's surface has been used to reflect light flashes sent out from a laser, which can emit very strong pulses of light. This has allowed the moon's distance to be frequently determined with a precision unattainable by any of the time-honored astronomical methods. Fourth, the astronauts have collected evidence on the arrival of atomic particles from the sun, the so-called solar wind.

FIGURE 13–4. Colonel Aldrin walking on the moon. In the center is seismic apparatus that he set up, and the lunar module is on the right. Courtesy of the National Aeronautics and Space Administration.

When it comes to sending probes to other planets within the solar system, the technical problems are greater because of the much larger distances involved. But in spite of these problems, a probe was launched to Venus in February 1961, and as early as December 1962 the United States probe Mariner 2 passed within 22,000 miles of Venus and sent back evidence indicating that the planet was very hot with a surface temperature of about 340°C, about three and a half times that of boiling water. This was not expected, as astronomers had thought the temperature was no more than 50°C. Probes launched by the Russians—Venus IV in 1967 and Venus V and VI in 1969—have been lowered by parachute through the very thick atmosphere of the planet and have confirmed that its surface is very hot; the temperature is now believed to be as high as 400°C, and the pressure of the planet's atmosphere near the surface is at least eighty times that on earth. Because Venus has a very cloudy surface, it has been impossible to use a telescope to determine how long it takes to spin on its axis, but space probes have helped radar (see Chapter 12) to give the value now accepted.

So far, no soft-landings have been made on Mars, and the information space probes have been able to give is confined to what they can detect as they fly past within a few thousand miles. In spite of this restriction the American Mariner 4, launched in 1964, took photographs that surprised astronomers because they showed that the surface of Mars had a number of craters. Mariner 4 passed within 6,118 miles of the planet, but in 1969 Mariners 6 and 7 went to within some 2,000 miles and showed that the craters are really numerous; the Martian surface is rather like that of the moon. The canals that some late nineteenth- and early twentieth-centuries astronomers believed they had observed certainly do not exist, and it seems probable that they were an optical illusion caused, perhaps, by many small craters linked together in a line. Mariners 6 and 7 also showed that Mars' atmosphere is composed mainly of carbon dioxide and is thinner than anticipated; in fact it is about as thin as our own atmosphere at twenty miles. When we remember that explorers need breathing equipment on Mount Everest, which is only five miles high, we realize that the Martian atmosphere is very thin indeed. This means that it has little power to stop x-rays and short ultraviolet radiation from reaching the surface, so life does not seem as likely on the planet as was hoped even a few years ago.

From this brief outline of the use of rockets and space probes it is easy to get a clear picture of how important they are if astronomy is to continue to expand, and our knowledge of the forces at work both locally and in the depths of space is to grow. Space research is expensive but recent developments make it seem clear that the expense will drop, especially when a manned orbiting observatory is launched in the early 1970's as the United States plans to do. Perhaps future historians will look back on the technical breakthrough that allowed space probes to become a practical way of observing the universe as an advance as important as the invention of the telescope.

14 ✳ The Age and Origin of the Universe

Now we have new ways of observing the universe; we are no longer limited to what can be done from earth, but can launch instruments into space; radiation that was once locked away by the atmosphere can be reached, and we can gather more facts and make measurements where earlier we had to guess. We also have better ways of processing our observations; they can be put on to punched tape and fed to computers, and there is no need to analyze each one separately. There was an old saying that for every night spent at the telescope, it took a week to deal with the evidence; this is no longer true. But though we have come a long way, there are still plenty of questions to be answered.

We cannot yet give a definite figure for the age of the universe, or even be certain of the earth's age, nor do we know which, if any, of the present theories of the origin of the universe is right. We can give figures, though, for the ages of the stars and make calculated guesses about the galaxies, and this is a great improvement on the first attempts, where little more could be done than place the beginning way back before historical events that were common knowledge. This satisfied earlier societies, for the mechanistic attitude to the universe only grew up in western Europe with the spread of the Renaissance during the sixteenth century. Prior to this, the outlook was different because the universe was seen as a symbol of divine creation rather than a piece of elaborate machinery, and if people tried to compute the exact motions of the planets, this was a way of seeing beneath outward appearances into the mind of the gods. But when the scientific attitude became stronger and old ideas were questioned, there was a desire to find the precise age of the universe. One of the first attempts was made by James Ussher, an early seventeenth-century Protestant archbishop of Armagh. He based his computations on the generations of patriarchs given in the Old Testament and came to the conclusion that the world was created in 4004 B.C. This seems a ridiculously small figure today, but when Ussher proposed it it was taken seriously, and accepted for a long time—indeed it is found in some bibles published during the 1850's.

But Ussher's time scale was soon challenged by Halley, who broke new ground and sought physical evidence of the earth's age, instead of turning back to the scriptures. Halley's idea, which he put forward in 1694, was that the age of the earth could be calculated by making regular measurements of the salinity of the oceans. He thought that as time passed the salinity would increase because of evaporation, and that once the rate was discovered one could work backward and compute how long ago the water was almost fresh. Halley did not work out any precise figure, but he was able to show that the earth was far older than Ussher's calculations indicated. He was accused of impiety for his pains, and it was a century before any other scientific attempt was made to assess the earth's age.

A new figure was proposed by the geologist James Hutton in 1795 in *Theory of the Earth*. Through his interest in farming Hutton became curious about the nature of rocks and minerals that were uncovered during plowing, and this set him thinking about the formation of the earth's surface features. He knew about earlier theories of the stratification of rocks, but he could not accept that all the features he saw were formed by sudden catastrophes, which current view fitted in well with Ussher's date of 4004 B.C. In place of this Hutton proposed that surface features were formed slowly but inexorably over a much longer time scale. He thought that changes were caused by erosion and that many rocks were formed by heat below the earth's crust, and though there were obviously some catastrophes, such as earthquakes and tidal waves, these were exceptions; the alteration of surface features was generally a slow uniform process, and Hutton's theory was called uniformitarianism.

Hutton's ideas were confirmed in 1830 when Charles Lyell published his *Principles of Geology* in which he brought together all kinds of evidence to support the long time scale and discredited the concept of catastrophes as a major influence. Further support came ten years later from the Swiss naturalist Louis Agassiz's comparison of fossil and modern fish and work on the formation and behavior of glaciers, which led him to suggest that the earth had been through a number of ice ages. Meanwhile other naturalists were changing their views and coming closer to the idea of evolution of plants and animals. In 1859 Charles Darwin published his theory of evolution in his famous book *On the Origin of Species*, which raised a storm of protest. A few of Darwin's critics were scientists who, though they had accepted uniformitarianism, nevertheless felt that his theory would make the time scale far too long. Gradually, however, the theory of evolution was accepted, and with it an age for the earth measured in at least hundreds of millions of years. If the earth was indeed as old as this, then it seemed to the late nineteenth-century astronomer that though the universe must be older still, its age was probably only two or three times greater.

These figures were soon to be revised when new calculations were made at the beginning of the twentieth century, owing to experiments on radioactivity. Working in Paris, Pierre and Marie Curie began to investigate the way in which some atoms spontaneously broke up, disintegrating on their own without any outside assistance. They discovered radium and found that when its atoms disintegrated they changed from radium atoms into atoms of particular kind, or isotope, of lead. Radium was found in some of the earth's rocks and inasmuch as the Curies had measured how quickly a collection of radium atoms disintegrated, it became possible to compute the age of a rock from the amount of the lead isotope it contained compared with the quantity of radium still in it. This particular method of radioactive dating (other more precise methods have been developed since) gave what seemed to be reasonably definite evidence that the earth had possessed a solid crust for at least 1,500 million years.

Astronomers were now stimulated to consider what figures they could derive for the age of the universe from celestial measurements, and from about 1910 onward they began to make some careful investigations. But to see how important these were we must first look at how the astronomer thought the earth and the universe had started. In the beginning, Western man had just taken the biblical account of creation from chaos as true, and had not inquired any deeper. The first scientific views about this did not arrive until 1755 when the German philosopher Immanuel Kant suggested that the solar system had begun as a shapeless nebula. This he believed to have rotated and shrunk until it became a flattish disk and then, on shrinking further, had formed into the sun and planets. This was part of Kant's earlier writings and was somewhat fanciful—in the same book he speaks of all the planets being inhabited, with the inhabitants proving morally better the farther one goes from the sun. This was hardly a scientific judgment, but at least his idea of the actual formation of the solar system made no use of supernatural interference—it was a purely mechanical suggestion and, as such, broke new ground.

Kant's idea was followed forty years later by a very similar theory, but this time it had a sounder backing, and its proposer was the French astronomer Pierre Laplace. It seems that Laplace did not know of Kant's theory, but even though he was a mathematician of considerable ability, who spent much of his working life applying Newton's theory of gravitation to every detail of the motions of the planets and of the moon, he did not work out the full mathematical consequences of his hypothesis. At this time, the creation of the universe was thought to be too speculative to bear such treatment, but nevertheless Laplace did go into more detail than Kant had done. He suggested that as the nebula rotated and contracted, its speed of rotation would increase until the outer parts were moving too quickly for the nebula's gravitation to hold them. A ring of material would then break off and split into fragments that would then join together to make a planet. The remainder of the nebula would shrink further, rotate faster, and another planet would form. This would be repeated until only enough material to form a stable sun was left at the center.

Such a way of forming the planets would occupy considerable

time and even if one accepted no more than a few hundreds of millions of years as a suitable figure, a big problem arose. How did the sun keep shining for so long? If it just burned away to give light and heat, it would long since have gone out, so its sources of energy must be replaced in some way.

In 1848 Julius Mayer, a German physician, made proposals that he thought might help to solve the question. He drew on recent astronomical research that had shown that meteors were part of the solar system, and on some physical research by James Joule that had made it clear that heat could be generated by friction; he combined these ideas and suggested that the sun replenished its energy from meteors falling into it. This was an ingenious hypothesis, but when Mayer calculated how much meteoric material would be required, he came up with the awkward result that the sun must absorb twice the moon's weight each year. If all this material remained in the sun after it had given up its energy, the sun would grow increasingly heavier and this would alter its gravitational pull, there would be changes in the motions of the planets, and, as far as the earth was concerned, the year ought to become less by a fraction of a second. Because no such changes could be detected, Mayer suggested that some of the sun's material itself became used up in generating heat. The significance of this suggestion of a change of matter into energy, which has played so large a part in theory and in practice (atomic energy) in modern times was not appreciated at the time. Indeed, the famous physicist William Thomson scorned the idea.

A new hypothesis came a few years later, in 1854, when the German scientist Hermann Helmholtz suggested that much of the sun's energy came from a contraction under gravity, so that material did not have to come from outside. Yet even this was not enough: Helmholtz himself calculated a lifetime of 22 million years, whereas others arrived at a rather smaller figure. All fell short of what the geologists demanded. Almost the whole scientific world seemed attracted by this particular conundrum. William Siemens, an expert in furnace design, suggested looking on the sun as a giant furnace, that sucked in material from space as it rotated, consumed it, and then ejected it like ash. But a full investigation by Thomson proved that even this, as well as other views, would not work out satisfactorily in detail, and for a time the subject was left unsolved.

As the years went by, the geological time scale for the age of the earth grew, and in 1887 Norman Lockyer produced a new meteoric theory far wider in implication than Mayer's. He suggested that because by then it had been found that meteors not only orbited the sun, but were also debris of comets, the accumulation of meteoric material over the ages must be considerable. Lockyer then argued that it was probable that there was a lot of meteoric material distributed throughout the entire universe, and concluded that "all self-luminous bodies in the celestial space are composed either of swarms of meteorites or of masses of meteoric vapour produced by heat." This meant that the whole universe of stars, nebulae, and planets had been formed from meteoric material.

Lockyer extended the meteor theory to the whole universe because Laplace's nebular hypothesis had become unpopular. On detailed mathematical analysis it was found the angular motion of the solar system given by Laplace's theory did not agree with what was observed. But Lockyer's views did not find much favor either, and in 1900 two Americans in Chicago, the geologist Thomas Chamberlin and the astronomer Forest Moulton, put forward a really novel proposal. They suggested that another star had once approached close to the sun and that gravitational forces between the two stars had torn out material from each. The result was that the sun had a cloud of gaseous material around it, which then condensed to form small solid lumps—planetesimals. The planetesimals subsequently joined together to compose the planets.

Chamberlin and Moulton's planetesimal hypothesis attracted interest. In England in 1916 James Jeans examined it in detail, but finally decided that there was no real reason why the planetesimals should join together to form the planets. He suggested a tidal theory, which posited the close approach of another star, but not that material would be torn out. He calculated that gravity would cause both stars to become oval and that only then would one star's material break up into lumps, each lump being large enough to form a planet; the planetesimal stage was no longer required. But this idea was only a first approximation, based on the idea that the gas inside both stars was evenly distributed from the surface down to the center. A second calculation, assuming the sun to have a dense central region, led to a slightly different state of affairs where material was torn out by the passing star. This material, Jeans computed, would fall back into the sun unless it was sufficiently large for its own gravity to keep it together as a separate lump; it would be sausage-shaped and lie with one end close to the sun, and the other toward the passing star.

Jeans's theory of the formation of the solar system held the field for some twenty years, but at the end of World War II the German astronomer C. F. von Weizsäcker revived the nebular hypothesis. The revival was possible because of the research into the generation of stellar energy (see Chapter 8) made from 1938 onward, from which it had become evident that hydrogen was the most abundant element in the universe. Von Weizsäcker thought that this materially altered the situation and, in 1945, suggested that if the nebula was composed mainly of hydrogen, or hydrogen and some other very light elements, then the planets could be formed not from rings of material but by whirlpools themselves coming from the rotation of the nebula. Three years later, in 1948, Frank Whipple of Harvard Observatory proposed a variation in which the sun condensed from a huge slowly rotating dust cloud, and then captured another more rapidly rotating dust cloud, which formed planets.

Laplace's theory was not the only one about the formation of the solar system to be revived in new guises, for in 1935 Russell suggested that perhaps the sun was once a member of a binary system. He supposed that a third star had come close to the sun's companion and whisked it away by a gravitational pull, with the result that a cigar-shaped lump was torn off and formed the planets. A full investigation by Raymond Lyttleton of Cambridge University showed that Russell's theory would not work, but in 1941 he found that the situation would be different if the sun had been a member of a triple star system instead of a binary one. Lyttleton's idea was that the triple star system consisted of a close binary system and the sun. Then, as this triple system moved in orbit round the center of the galaxy, Lyttleton supposed it would sweep up dust and gas, and the stars would become larger and the binary pair would move closer to each other until in the end they became one star. The speed at which this star would spin would be so great that it would break up again. Its two pieces would rush off into space leaving a lump of gas to form the planets as it moved in closer to the sun. Once again, difficulties have shown up on further study, and astronomers have been seeking alternatives. Fred Hoyle at Cambridge developed Lyttleton's and Russell's theories in 1944 by analyzing what would happen if the sun were a member of a binary system in which its companion became a supernova and suddenly exploded. His calculations showed that though the supernova would carry away some gas with it, enough would be left behind with the sun for the planets to form the solar system. This idea has one important advantage in that it makes available to the planets the atoms of such heavy substances as iron, lead, and uranium, substances that would be very rare or even nonexistent if the universe only began with hydrogen gas, the simplest and lightest of all atoms. These heavier atoms are built up deep inside stars and would be released in a supernova explosion. With the American atomic scientist William Fowler, Hoyle has since gone into this in greater detail, and the theory seems promising. Yet in spite of everything, we still cannot be sure either of when the earth was formed or precisely how. What does seem to be certain is that the sun came before the earth and played an important part in the origin of the solar system. The planets formed out of dust or gas—or both—but the details still elude us, or at least there is no one explanation that all astronomers will agree is the correct one. Again, there is doubt about the time scale but, by and large, a figure of some 4,500 million years is generally accepted.

But what of the age of the universe? This is a far more difficult problem, and figures have varied widely since Lockyer's guess of a couple of thousand million years. This was bold enough when he made it, but as it soon became clear that the earth was older than this, there was need of a new time scale for the universe. What could be done to make a new assessment? What astronomical evidence was there that could provide a guide line? An obvious clue should come from the ages of the stars, for clearly the age of the universe must be greater than the age of the stars it contains. The trouble here is that at the beginning of the twentieth century no one was at all certain how quickly stars aged. Kapteyn's studies of the stellar motions began to shed new light on this subject. In 1904 Kapteyn showed by applying statistics to the observed motions of the stars, that far from traveling at random, they were really streaming along in particular directions. We now know that Kapteyn's studies were the first sound evidence for the rotation of the whole of our galaxy, but their importance in the struggle to determine the age of the universe lay in the use made of them by James Jeans. Jeans decided to examine open clusters of stars— such clusters as the stars of The Dipper in which all but the brightest (alpha) are moving in the same direction. He realized that as the cluster moved through space, each star in it would be affected by three things: (1) the gravitational pull of the rest of the stars in the galaxy, (2) the gravitational pull of the other stars in the cluster, and (3) the effects of nearby stars not connected with the cluster. He saw, too, that this last effect would in due course lead to the dispersal of the cluster of stars. Jeans computed that this would take a long time to happen and that a figure of 1 million million years was about the correct value. Another study that he made, this time of binary stars and the way these would become parted after a long time owing to the gravitational pull of nearby stars, gave a similar result. So by 1928 Jeans was able to calculate that the age of the universe must be something in excess of 1 million million years, and possibly as great as 10 million million years. In this he disagreed with Eddington who did not accept some of Jeans's reasoning, and believed in a time scale that was about a thousand times shorter. Jeans's long time scale for the universe was adopted by many until the importance of Hubble's velocity-distance measurements of the movement of the galaxies became well known (see Chapter 11). If Hubble's results were worked backward, calculating how long it would take for the galaxies to come together again, the time came out at 2,000 million years, a figure far smaller than the one obtained by Jeans. Obviously something was wrong somewhere; as it turned out, both values were wide of the mark.

Jan Oort and Bertil Lindblad showed the flaw in Jeans's arguments by proving that the stars do not all orbit round the center of the galaxy at the same speed, but that those near the center move more slowly than those nearer the edge. In consequence, the stars in an open cluster would suffer more irregular gravitational effects than Jeans had calculated, so his long time scale was far too long. But the short time scale seemed far too short and, though it was necessary to move from Jeans's protracted age for the universe, the question was where to go next. Only gradually was it possible to expand the lower value, but from 1952 when Baade showed that earlier estimates of galactic distances were too small (see Chapter 11), astronomers have generally accepted the idea that the universe is not substantially older than some 10,000 million years —at least a hundred times shorter than Jeans's figures, which are now completely cast aside.

Yet it would be wrong if we left the matter here, for although Jeans's long time scale was rejected in 1953, the question of the origin of the universe was far from being solved by a majority acceptance of 10,000 million years. By 1958 the astronomical world was divided into two camps when it came to considering how the universe began. In 1933 and 1934 Lemaître had put forward a theory to account for the original formation of the universe. Essentially this considered a giant lump of material—a sort of superatom—that was unstable because of its complexity, in just the same way as large atoms of such substances as uranium are unstable. The superatom exploded and its material rushed outward, breaking up into the kind of atoms we now observe in the universe. It is from these atoms that the galaxies have condensed as general expansion continued. Lemaître did not think that the force of the original explosion would be sufficient to keep the expansion going because the gravitation of the galaxies would tend to slow things down. To counteract this deceleration he brought in the relativity concept of cosmical repulsion (see Chapter 11), which takes over from gravitation at great distances.

Another theory starting with a superatom, but this time making no use of cosmical repulsion was first proposed in 1939, and then in more detail in 1948, by George Gamow, R. A. Alpher, and Hans Bethe, so that it is sometimes known as the alpha-beta-gamma theory (Alpher, Bethe, Gamow). They assumed that the universe began as an immense lump of very hot atomic particles known as neutrons. Some of the neutrons split up into other simple atomic particles (protons and electrons), and some of the protons joined the neutrons to form the central cores of heavy atoms. They worked out that the heavy atoms would be constructed within the first half-hour of the universe's existence, but later studies showed that this was being a little too optimistic about the powers of the theory. In the end they have had to be content to assume only the build up of the atoms of the two lightest substances—hydrogen and helium—in the very early stages. Gamow had no need to call on cosmical repulsion, because the build up of more complex atoms from simple atomic particles (nuclear fusion) is a much more powerful force than the breakdown of complex atoms (nuclear fission) which Lemaître envisaged. Because Lemaître's and Gamow's theories both start with an exploding superatom, they are known as big bang theories.

Both the big bang theories ignore one important point: where did the superatom come from? This is a question that astronomers cannot answer, though there are two theories of the origin of the universe that do away with the need to ask it. One, the steady state theory, claims that the universe always appears to be the same. It was suggested by Herman Bondi, Fred Hoyle, and Thomas Gold at Cambridge University in 1948. We know that stars are formed, live their lives, and die; that some explode; that the universe is expanding. But what Bondi, Hoyle, and Gold claimed was that if a large enough sample of the universe is taken, then stars that are being born will compensate for those that are dying. (See Figure 14–1.) In fact, though minor changes will go on, there will be the same general picture all the time; though galaxies may move away owing to expansion new ones will form to take their place. In their theory, if you shot a movie of any part of the universe you would see small alterations, but if you compared movies shot now, 1 million years ago, and 1 million years ahead, you would be unable to decide which was which just by looking at them.

In the steady state theory, the universe never had a beginning and it will never have an end. Different objects in it will have their own lives and it is quite meaningful to talk of the age of the earth,

FIGURE 14–1. The Lagoon nebula in Sagittarius (The Archer), M 8. The little dark patches may be material condensing to form stars. Courtesy of the Lick Observatory.

or the age of the sun, or the age of our galaxy. But it is nonsense to talk of the age of the universe. However, it still leaves open the question of where fresh galaxies come from. If galaxies are moving away because of expansion, and new ones must form, how can this happen? They overcame this difficulty by proposing the idea of the continuous creation of matter, Hoyle actually suggesting that the central parts of hydrogen atoms (the simplest atoms) are continually forming in the universe. These then come together and form gas clouds and then galaxies.

Astronomers have been very divided in their opinions about the steady state theory. It offers some advantages but some disadvantages too, and many have been disturbed by the idea of the continuous creation of the nuclei of hydrogen atoms, for in Hoyle's calculations the number required to replace the material lost by expansion is too small to be observable, so that on this count the theory can neither be proved nor disproved by observations. Nevertheless we shall see in the next chapter that there are observational arguments to support it in a modified form.

Another of the possible models of the universe is to take an expanding universe (such as is observed) and to remove the idea of cosmical repulsion. What then happens—in theory—is that a time comes when the energy of expansion is used up, and the universe begins to shrink. It goes on shrinking until everything comes very close together and a great deal of heat is generated; then the universe begins to expand again. This expansion-shrinking-expansion goes on forever, so that such an oscillating universe has no more beginning or end than the steady state theory. The first example of an oscillating universe was given in 1922 by the German mathematician A. Friedmann, but it is only now, when we step from the pages of history into the astronomical world of today that the theory is being carefully examined.

15 ✳ Looking to the Future

The sketch we have drawn of the history of astronomy has shown that some discoveries were more important than others and that a few have reshaped our entire outlook on the universe. In this chapter we shall try to use the facts of history to help us put the discoveries of modern research into their proper place, so that we do not make the mistake of thinking them world-shattering when they are not and, on the other hand, do not ignore what is important because we cannot see its significance. In fact, our historical knowledge can help put recent events into their proper perspective. We shall also try to do something else that is a little more difficult—to see whether we can guess how things will go in the future, basing our guesses on the lessons we can draw from previous history and from present results.

Let us begin with a relatively simple but important matter—the first landing of astronauts on the moon. There are some things about this that are fairly straightforward. First, the successful landing and return means that, from a technical point of view, space exploration has now arrived, and there is enough engineering knowledge to take men farther into space—at least to Mars and, possibly, as far as Jupiter. Second, the success gives yet another confirmation of Newton's theories (for the difference between these and Einstein's theory are too small to show up on so short a trip): gravity on the moon was what calculation had led us to expect, and all the orbiting maneuvers went just as the computers had programmed. Third, the lunar rock samples were scooped up without much difficulty, photographs were taken, even stereoscopic close-ups of the lunar surface that showed pieces of rock as small as .002 inch. Experimental equipment was left on the moon and has continued to behave just as it should long after the astronauts left.

All this has important implications for present and future astronomy. As far as present research is concerned, the moon landing has very strongly confirmed the astronomer's belief that only the chemical elements we know on earth, and no others, are to be found elsewhere in the universe. Ever since the Greek idea that the stars were made of a special celestial substance was rejected during the middle of the seventeenth century, it has really been a matter of faith, though things did change a little in the nineteenth century because of two developments: the invention of the spectroscope and the discovery that pieces of meteoric rock had originated in the solar system. Neither of these was proof that there were no more elements, but it made it seem more likely that the same substances we know appear throughout the universe; but this belief received a setback almost immediately from the researches of William Huggins (see Chapter 9). About 1874, when he was studying the nebulae, he found that on examining the gas clouds in the galaxy, there were lines in the part of the spectrum that he could not identify with any known lines in the laboratory. Others tried to make an identification and failed, and it looked very much as though everyone was wrong, and there might be at least one unfamiliar element out in space. The situation altered in 1895 when, using clues obtained from the spectrum of the sun, Ramsay managed to isolate helium in the laboratory. Yet still Huggins' green lines remained, and a great number of astronomers became convinced that they represented some unknown element, which they called nebulium.

The problem of nebulium was not solved until 1927, about half a century after its discovery, when the American astrophysicist Ira S. Bowen found that when he put oxygen and nitrogen in an almost completely evacuated glass tube and then electrified them very strongly, they emitted the green lines of nebulium. In other words, the lines were caused by chemical elements only too well known on earth, but when they were under the most unusual conditions. Now another piece of evidence has come from samples of moon dust and pieces of lunar rock, for these can be handled, analyzed, and compared directly with terrestrial rocks. The results show differences—the rock basalt, familiar on earth and present in lava flows, is found on the moon but the chemical composition is not quite the same; but the important thing is that no elements are present in lunar rock that are unknown to the chemist. And since the first moon landing captured the imagination of scientists as well as the public, it is a more dramatic and perhaps more effective confirmation.

A second and equally important aspect of the first moon landing is that active experiments can now be conducted on the universe, at the moment only a small nearby piece of universe, but another body out in space all the same. The experiments carried out there by the astronauts were different from the kind of passive analysis that astronomers can do back on earth with moonlight, or even the experiments that can be conducted by remote control using such unmanned probes as the Surveyor space craft. To go out in space and perform experiments on celestial bodies has always been the astronomers' dream, but until the success of Apollo 11 this was no more than a hope. Clearly, the landing is of the utmost importance historically, and it opens a new chapter in man's investigation of the universe, in which one of the next steps will be establishing bases in space and on the moon, complete with radio, optical, ultraviolet, and x-ray telescopes (see Figure 15–1).

Next let us go further than the moon and turn our attention to another recent development in astronomy—the discovery of pulsars. Like some previous discoveries in science, such as x-rays and penicillin, pulsars were found by chance. If Jocelyn Bell and Anthony Hewish had not watched out for repeats of the first strange pulses they had detected and not been bold enough to eliminate the likelihood of messages from little green men (see Chapter 12), it would not have been possible to investigate the objects so successfully. As it was, constant attention to detail gave enough information for optical astronomers to pick up the pulsar in the Crab nebula and, at Lick Observatory, using the latest electronic techniques and a television camera to photograph it. So far the discovery has followed the usual scientific pattern, but when it comes to an explanation of what a pulsar is, things are more difficult.

The pulsars are characterized by their regular emission of pulses, each of which lasts no more than a fraction of a second, and it is the regularity of these pulses that is so remarkable. We can get a clue to their nature if we can find how far away they are, and though one has been identified as a flashing star in the Crab

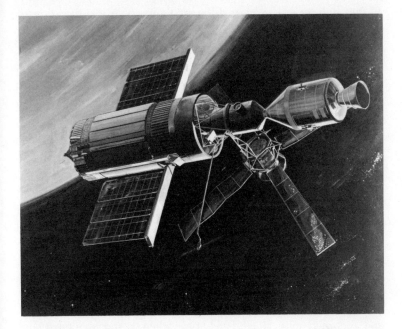

FIGURE 15–1. An artist's impression of the projected Saturn 5 space workshop, which has telescopes for observing space (center) and is a docking station for space probes going to Mars and perhaps beyond. Courtesy of the McDonnell Douglas Aeronautics Company.

nebula—which gives it an approximate distance of 3,500 light-years—this by itself does not mean that the rest may not be nearer. The one thing that seems certain is that they do lie inside the galaxy. In an attempt to assess distance, workers at Cambridge, England, devised a method of measurement based on the fact that radio pulses spread out in the space between the stars. This space is not empty and the interstellar dust affects the speed at which the radio waves travel, making the shorter wavelengths slow down more than the longer, so that by measuring the pulses at different radio wavelengths, and assuming a certain average density for the dust, they found distances of between 100 and 5,000 light-years. This seemed to be fine, until a pulsar distance estimated by another method, based on the way spiral arms in the galaxy absorb radio waves, gave distances of no fewer than about 13,000 light-years—a very different figure. The identification of one pulsar with the Crab nebula (see Figure 15–2) makes the first method seem to be the correct one, but why is the other in error? Is our evidence about the spiral arms of the galaxy wrong, or were the measurements faulty? Or is the Crab nebula pulsar an exception as far as distance is concerned?

The question of distance is crucial because astronomers want to account for these strange objects, and the explanation will depend on how strongly they emit radiation. But strength or radiation can only be calculated when we are certain of distance, and there is still plenty of room for doubt. Nevertheless various explanations have been given and, in Chapter 12, we noted the suggestion that a pulsar is essentially a neutron star spinning at high speed. On the face of it, this appears a good way to account for the facts, but can the matter be left there? From a glance at history, where observations have been made of objects previously unknown, it may be possible to obtain a clue, and one example immediately comes to mind—the supernova. Supernovae have been observed and recorded ever since 1054 A.D., when a star exploded and resulted in what we now observe as the Crab nebula; yet neither in 1054, nor in 1572, nor in 1604 was it realized that they were exploding stars. In 1572 when an object appeared in Cassiopeia, Tycho observed it and noted that it was as bright as Venus for a while, and by his observations showed that it lay well beyond the moon. He thought it must be a new star and, in 1604, when Kepler observed a supernova in Ophiuchus, he too put it down to the sudden appearance of a newly created star. This was agreeable to most astronomers because, though they still clung to the belief that the heavens were changeless, a new star seemed to them evidence of a new act of divine creation and this would overrule any limitations imposed on the universe. Only when the legacy of Greek thought had been disposed of was it possible to think of a

FIGURE 15–2. The Crab nebula, so called because early drawings made in only some of its light (to which the eye is sensitive) gave it the appearance of a hermit crab. It is the remains of a supernova explosion and contains a pulsar. Courtesy of the Hale Observatories.

different explanation, but not until after William Herschel had discovered variable stars (see Chapter 6) during the late eighteenth century could a star's change in brightness be considered.

In the case of pulsars, the explanation that these are rapidly rotating neutron stars uses a star type with which the twentieth-century astronomer is familiar in theory, if not in practice. This kind of approach is a logical first step, for when any new or strange object is observed, it is better to see whether it is like what is already known rather than invent something totally new. However when such possibilities have been exhausted history shows that it is then time to propose something novel. So far as pulsars are concerned, this is what history does indicate, and more recent observations make it seem to be a better course. Australian radio astronomers have found a pulsar in the southern Milky Way that gives pulses that appear with perfect regularity but vary in strength without any apparently recognizable pattern and an American group has found variations every three months in the strength of the Crab pulsar. A straightforward rotating star would scarcely account for these observations, and though an orbiting planet or a binary star system have been proposed, neither seem good enough. Possibly, a pulsar is some new kind of pulsating variable but, whatever it turns out to be, we are left with the impression that it is most likely to be a type of object at present unknown to the astronomer.

But of all the questions facing astronomy today, the most difficult to answer and the most important to decide is the nature of quasars; observationally they bristle with difficulties and seem to give astronomers a whole host of inconsistent answers. They show both dark as well as bright lines in their spectra, so they would seem to be stars surrounded by gas, but they are clearly not stars because their radio energy is far too great for any star with which we are familiar. Yet they are known to be small and not so very much larger than stars, both because they appear as dots on a photograph and because their radiation varies fairly rapidly, sometimes over a period as short as a week. Such variation could not be detected if the bodies are large because light takes time to travel, and a quasar varying its light every week cannot be larger in diameter than one light-week (1/52 light-year), otherwise the light from the rear would be different in intensity from the light at the front and the variation would be so muddled by a mixture of different brightnesses that we should not detect it. As a result it is found that many quasars must not be larger than about 1/3 light-year, and some seem to be as small as 1/52 light-year or about 100,000 million miles; this rules out the possibility of a quasar being an ordinary galaxy. Another inconsistency has arisen because, though most quasars radiate vast amount of radio energy, some emit so little that they can hardly be detected by radio telescopes.

All quasars have immense red shifts, in some cases greater than those of any galaxy and, in the argument of greater red shift with increasing distance (see Chapter 10) some astronomers have taken these to mean that quasars are very distant objects. Astronomers believe this cosmological interpretation bolsters the big bang theory. Their argument is that as quasars are so distant, what we are observing are objects as they were 6,000 million, 7,000 million, or 8,000 million years ago (because their light has taken this time to reach us); in other words, in observing quasars we are seeing the universe as it was close to the time of creation. This could mean that quasars are galaxies in a very early stage of formation, and probably shows what they once looked like when incompletely condensed. This is a straightforward and simple explanation, but it does bring the astronomer face to face with a very difficult problem: if quasars are really distant, then though they are small they emit as much energy as at least ten whole galaxies of stars. And the trouble has been aggravated recently after observations from space have made it clear that, energetic though a quasar may be at radio wavelengths, it is an even more energetic emitter of x-rays and short wave ultraviolet. No known, or calculated, process of nuclear fusion can give such energy, so the cosmological interpretation does encounter a serious difficulty. One way out has been proposed by Greenstein and Schmidt, who think that a quasar may be a cloud of very hot electrified gas with a very dense center. Atomic particles would be accelerated to immense speeds inside such a gas and radiate very strongly just because of this motion. This explanation might make sense, they believe, not only of both bright and dark lines in the spectrum but also of another unusual feature associated with quasars—the multiple red shift. A close examination of a quasar spectrum shows that the bright lines have two different shifts, and a surrounding cloud of gas could give this effect if it was in violent motion, because some parts would be moving toward the observer and others away, even though the cloud as a whole is receding quickly.

But is the cosmological interpretation of the red shift correct? History gives us reason to doubt its truth, as we shall see if we glance back at the discovery of the red shift of galaxies. When the shift was first found, the distances of the galaxies were unknown and it was assumed that they were inside the galaxy. An orbital motion round the center of the galaxy would then give the red shift, though it would also cause a motion across the sky. It was thought that this sideways motion would be detected when enough time had elapsed to make it large enough to be observed.

In the years since the discovery of the expansion of the universe, some astronomers have certainly questioned whether the Doppler-Fizeau explanation of the red shift is right, but though at least three careful attempts have been made to find another cause, no one has been satisfied with them. The first, which suggested the red shift is no more than an optical illusion owing to distance, was rejected almost at once and another, supposing that light lost energy on its way across space, was unable to account for more than a tiny fraction of the shift. The third came in a modified theory of relativity suggested by Edward Milne, in which radiation worked on a different time scale from events on earth and so gave an apparent red shift for very distant objects. But this has been considered too theoretical, because there is no observational evidence to support it. Yet there are three possible alternatives to the cosmological interpretation of quasars: (1) that the red shifts are an effect to be expected from relativity theory; (2) that they result from high velocity but are not linked with great distance, and (3) that they appear because of some quite unknown reason.

The relativity effect arises because a large body with a strong gravitational pull attracts radiation strongly, just as it attracts other bodies. Light emitted from it will show a red shift, the amount of the shift depending on the strength of the gravitational field, and for a shift of the degree observed in quasars, the bodies would need a very powerful field, which means that they must be immensely massive. But if there are such massive bodies lying within the galaxy, we should expect to find them causing some other gravitational effects—distorting the way the stars move, for instance. Even though no such results have been observed, it could still be argued that an Einstein shift is possible, provided quasars lie outside our galaxy, embedded in a gas that is most concentrated at the center of the collection. Calculations show that such clusters could give most of the observed effects, but they would have to lie between 30 million light-years and 300 million light-years if their gravitational effects on the galaxy are not to be noticeable. And still there remains the problem of how so massive a collection of bodies—about 10 billion times that of the sun—could possibly exist in so small a space. Nevertheless a gravitational red shift remains a possibility, even if not a very likely one.

But if we assume the red shift is a real velocity effect, that it is owing to motion directly away into space, we do not have to accept the cosmological explanation; it could be that quasars are the results of immense galactic explosions. If there had been such an explosion in the galaxy's central regions some time in the past, the quasar pieces coming in our direction would have moved past us, so that we should see them and all the rest of the resulting quasars with a red shift. Yet it is not very likely that our galaxy is the only one out of all the millions of galaxies known to undergo an explosion; it might help here if we could obtain evidence of galactic explosions in other parts of space. Recently fresh observations have been made of some galaxies that are strong emitters at radio wavelengths—radio galaxies—and in the case of the elliptical galaxy M 87 in Virgo, photographs show a jet of material being ejected. (See Figure 15–3.) The material is strongly radiating x-rays and short wave ultraviolet, and without doubt represents an immensely energetic explosion. Some other radio galaxies also show evidence of eruptions either because they seem to be splitting, as in the case of a galaxy in Centaurus, or throwing out lumps of material, so there is a growing amount of observational support for the explosive galaxy idea.

This explanation too comes up against a difficulty for, if explo-

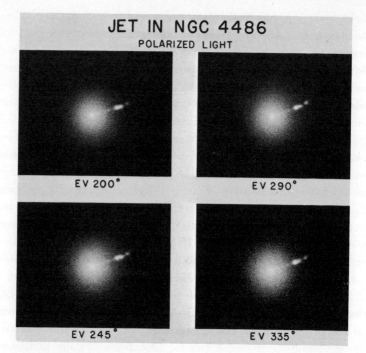

FIGURE 15–3. The elliptical galaxy in Virgo, M 87. Recent photographs, such as this one, show a jet of material being ejected from the center. Courtesy of the Hale Observatories.

sions are occurring in galaxies scattered all over space, and quasars are being thrown out, we should expect the quasars to be traveling in all directions, some toward us, some away from us, and some sideways. In consequence, we should observe some blue shifts as well as red shifts, but no quasar has been found with a blue shift. Nevertheless, the fact that not one blue shift has been discovered does not completely demolish the explanation, because it is possible that as quasars have been detected mainly from radio astronomical observations, we are making a special selection. This is not intentional but, because of the way quasars were originally found, it happens that our present list is drawn up with a bias. If we can find another way of detecting quasars, it may be that some will have blue shifts, for optical astronomers have already come across some strange objects that look like a cross between small galaxies and small quasars, and whose red shifts are smaller than they expected. And there is another point: many quasars are associated with galaxies that seem to be undergoing some kind of substantial upheaval. Indeed the association—closeness in space —is far too great to result from chance, and this is perhaps the strongest of all indications that quasars have something to do with galactic disturbances of one kind or another.

There is the possibility that quasars are some completely new and unknown object: something that looks like a star but is not one, something that might appear to be a tiny galaxy or a galaxy in formation, and is not. Certainly this idea cannot be ruled out. And if we want a historical parallel, we do not need to look further than the nineteenth-century struggle to solve the problem of the nebulae. To begin with (see Chapter 9), it seemed to be only a matter of deciding whether or not a really large telescope could resolve the hazy nebulous patches of light into separate stars, yet when Rosse built his seventy-two-inch reflector, it became clear that some of the objects appeared neither as hazy, ill-defined clouds nor as clusters of stars; they had a spiral shape. Surprisingly, Rosse and his colleagues did not realize that they had stumbled on a new kind of object—spiral galaxies—but were so concerned with the question of resolving power that all they did was conclude that all nebulae would one day be seen to be no more than conglomerations of separate stars. Their preoccupation with one problem restricted their imagination so severely that they could not think on a broader basis about what they had seen, even though spiral galaxies are unique and nothing like them had ever been observed before.

Today, certainly, more imagination is being used; astronomers are not sticking to only one idea for quasars, even if the majority favor the cosmological explanation because it supports the big bang theory, which is presently the favorite cosmology. Other possibilities are being carefully considered. We must recognize that here is something so unusual that it requires a new look at the universe and makes it necessary for us to realize that if quasars are not completely new objects, then at least they must be the result of some new and unexpected process of which galactic explosion is one possibility. And, of course, there is always the chance that what we classify as quasars may be more than one kind of object; such misunderstandings have happened in the past, as when all stars were thought to be the same and when no distinction was made between spiral and elliptical galaxies.

A new approach seems necessary not only to quasars, but also to the theories of the universe discussed in the last chapter. Here the astronomer must really use his imagination and be prepared for some basic rethinking because, from a historical point of view, the big bang theory appears to be a little too simple, and even the steady state theory cannot be expected to remain unchanged. The big bang was an obvious first step to take after the discovery of an expanding universe, but theories are not usually so straightforward and, as new observations arrive and more facts need to be fitted together, every hypothesis is broadened and becomes more complicated until, in the end, it has to be rejected or reconstructed. As far as the steady state theory goes, it was forced to suppose the appearance of fresh material all the time; it did this by assuming that the creation of atoms occurred evenly throughout the universe. This again was a first step, taken because the mathematics is easier if creation occurs everywhere and because there was no observational evidence to prove otherwise. In recent years the situation has changed and galactic explosions, such as the one observed in M 87, show that the strictly unchanging situation of the steady state universe is untenable; perhaps a modification is possible.

Already in the mid-1960's Hoyle produced a change to the steady state theory to account for radio observations made by Ryle at Cambridge, and by Australian radio astronomers. What they had observed was that there were more dim radio sources than bright ones in any given area of the sky and, assuming all the sources were radio galaxies and nearly of the same brightness, they concluded that there were more in the far distant parts of space. They then claimed that their observations strongly supported the big bang theories, because these state that in early times the galaxies were closer together because expansion had not been going on for so long, and observations of really distant objects shows the universe as it used to be.

However, as it turned out, the steady state theory was not

demolished by these observations because the radio astronomers could not be certain what the sources were they had observed and, even if they did happen to be galaxies, there was another interpretation of the results. It could be that the galaxy lies in a rather empty region of the universe where there are fewer bright sources than usual, and though this meant that supporters of the steady state theory had to admit that there might be differences in one part of space from another, they claimed that this did not prevent it being the same if one took a large sample that would incorporate our local area and the more distant regions observed by the radio astronomers. Another attack on the theory came in 1965 when a permanent background of radio radiation was found to exist in space; again those who favored the big bang theory saw this as evidence of the one-time existence of a hot gas that had now spread and cooled owing to the way the universe had expanded. Such a hot gas would be expected on the Gamow big bang theory, and the steady state seemed doomed. But further investigation has shown that the radio radiation is not of the correct kind and that whatever it may be owing to, a cooled gas is not the right answer; the steady state theory had another reprieve.

The discovery of galactic explosions has a further bearing on the steady state universe because in 1964 William McCrea of London University suggested that new atoms are created inside the central regions of galaxies, and not evenly throughout space. His idea was that in these very dense central regions atomic particles would break down under the great strain, and new particles would be formed, so that as a rule galaxies would tend to become more massive; occasionally, the new atoms would disturb them so much that they ejected matter from deep within their centers. The ejected material would then form the core of a new galaxy. This explanation covers the appearance of fresh galaxies to replace those moving away owing to expansion, and one can argue that in the case of M 87 we are actually seeing a new galaxy being born. Hoyle and a colleague, Jayant Narlikar, have again modified the steady state theory so that the creation of new matter no longer occurs all over space but in separate areas; they think it could be caused by the collapse of massive bodies inside or outside galaxies. Possibly quasars may be the cores of new galaxies, or even places where new atoms are being created, though in either case this means that they cannot be very distant.

With these modifications of the steady state universe we come to some radically new ideas, and this makes one feel that the astronomers developing them are following a course that history leads us to expect to be fruitful. But perhaps even the modified steady state theory is not imaginative enough, and Narlikar has suggested a bubble universe, where the universe considered as a whole is in a steady state, but in which there are thousands of explosive expanding bubbles. He assumes that we and the entire expanding universe that we observe form only one such bubble. Another possibility, favored by Robert Dicke of Princeton, is an oscillating universe (see Chapter 14), but to procure observational support for it we must one day be able to notice a slowing down in distant regions of space. No absolutely definite evidence of this is yet available, but that does not mean it never will be; until it is, this can be no more than a theoretical possibility.

By now it must be clear that we live in a revolutionary stage in the development of astronomy. Our ideas and horizons are being greatly expanded, and to a degree that has never happened before. The whole question of the nature of space is being reinvestigated, for it is now realized that the presence of dense objects, if that is what quasars are, may bring great distortions. It has even been suggested that space could be distorted as indicated in Figure 15–4, so that observers situated in one of the legs (A or B) of the

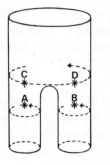

FIGURE 15–4. The pair of pants universe—an imaginative possibility.

pair of pants has a limit to the universe he can see because light beams bend over (as indicated by the dotted line). But an observer moving into a less densely populated part of space (C) will find light beams less strongly bent; he will see farther and will observe the appearance of new material at D that was invisible a moment before and will think that it had been created suddenly, because it will appear to come from nowhere! Perhaps the pair of pants space is rather farfetched, though it is mathematically feasible, but at least it underlines the point that we have had too many assumptions about space for too long. We still think of it in terms of everyday geometry and this is just as much a limitation to our understanding as failing to realize that quasars may be more than one kind of unknown object. It may be as difficult for us to think of space in quite new terms as it was for scientists 400 years ago to move from a finite geocentric universe to an infinite heliocentric one, but the lesson of history is that we must overcome our prejudice if we are to move on and bring closer the time when we have the real answer about our astounding universe.

Index